# BOTANIC GARDENS

## MODERN-DAY ARKS

# BOTANIC GARDENS
## MODERN-DAY ARKS

Sara Oldfield

The MIT Press
Cambridge, Massachusetts

# CONTENTS

# FOREWORD

Botanic gardens and conservation have been causes close to my heart since childhood. It is therefore an honour to write a foreword for this excellent and wide ranging look at the work of botanic gardens and BGCI, as it gives us a fascinating insight into not only the plants, but also the unsung heroes – the gardeners and botanists – who nourish and save them. The work of BGCI is to help conserve the plants of our earth. Not only do these provide humans and animals with their basic needs – food, shelter, warmth and medicine to name but a few of their values – they also play a major role in relation to climate change. BGCI is fundamentally important, successful and unique, run by a small dedicated team with headquarters based at the Royal Botanic Gardens, Kew.

This book will not only show you some spectacular plants but also give fascinating insights into botanic gardens and the extraordinary plants they protect. They need your help. Please help us to continue our work saving our planet and plant diversity which will, in turn, save our children's futures. You can do this by visiting gardens near you or visiting the BGCI website and supporting them directly. Please help spread the word to help fight plant extinction.

BETH ROTHSCHILD

**1** | **The plant extinction crisis**

Plants define the landscape of life. The colour in our gardens, the structure and function of ecosystems, the global supply of food and the air we breathe all depend on plants. But plant diversity is declining dramatically, threatening the loss of essential resources and the collapse of ecological stability. Scientists estimate that at least one-third of all flowering plant species are threatened with extinction. The impacts are most severe in the poorest parts of the world, which suffer most acutely from deforestation, desertification and famine, but ultimately the loss of plants and impoverishment of ecosystems will affect us all. With skilled scientists and horticulturists, and invaluable collections of living plants and reference material, the botanic garden community is uniquely placed to help tackle this crisis.

**Previous page:** Land clearance in Tanzania. Devastation of plant diversity in Africa is expected to increase as climate change adds to the pressures of removal of natural vegetation for agriculture and settlement. Conserving plants in botanic gardens provides options for ecological restoration.

Understanding plant diversity and the individual values of plants depends on studying, describing and cataloguing the different species – work that is undertaken by botanists collecting specimens in the rainforests or working in the world's herbaria. The process of cataloguing all the world's flora is not yet complete, but currently it is estimated that there are around 300,000 flowering plant species along with 12,000 ferns and 30,000 so-called lower plants. New species are still being described but this in no way compensates for the loss of plant species in the wild as part of the plant extinction crisis.

Some of the plants in trouble are naturally rare, confined, for example, to a single Mediterranean mountain top or small oceanic island, others are more widespread but under threat as a result of loss of habitat or over-harvesting for timber, food or medicine. Clearance of natural vegetation for agriculture, mining, urbanization and industry puts pressure on wild plant species and ultimately destroys the productivity of the land. Invasive species are another major threat – aggressive weed species are out-competing locally adapted native species in many of the world's ecosystems. On top of these threats, global climate change is compounding the problems for plants and

*Narcissus triandrus* subsp. *capax* is a rare daffodil growing in the wild only in Brittany. It has been rescued from near extinction by the botanic garden in Brest.

ultimately threatens human survival. Over the millennia plant species have adapted to changing climates or have migrated as a gradual response to temperature and rainfall changes, but now dramatic climatic changes are simply too rapid to allow plants to respond. Botanic gardens are finding a new role as modern-day arks, rescuing plants from the brink of extinction.

Dedicated botanists and gardeners are using specialist skills to try and ensure that plant species survive to inspire and delight. But we rely on plants for much more than decoration. It is not an exaggeration to say that all life depends on plants. Looking at our own diets, 30 main crops feed the world, but it is estimated that about 30,000 species are edible and about 7,000 have been cultivated or collected by humans for food at one time or another.

The work of botanic gardens in looking after the world's plants is immensely important. Plants have diversified to maximize productivity in different climatic and soil conditions around the world; they give ecosystems their structure and provide habitat for other species. With each plant species that becomes extinct, the insects and other animals that defend it are also threatened with extinction. But the diversity of plants doesn't just help to maintain the world's ecological stability, it also underpins our economic systems – providing the raw materials for agriculture, forestry, medicines and fuel.

## A SHARED RESPONSIBILITY

Looking after the world's biodiversity – animals, plants and ecosystems – is a shared responsibility. Governments, voluntary organizations and individuals all have a role to play. Plant diversity is championed by expert and amateur botanists and plant-lovers alike. There are dedicated individuals who devote their working lives to saving plant species from extinction, people who grew up with a fascination for plants, studied the scientific basis for their diversity, physiology and ecology and, if they were lucky, went on to make a career in botanical conservation. Usually the professional plant conservationists work in botanic gardens. Over the past thirty years or so the need for plant conservation has become increasingly recognized and new measures are constantly being developed to save the world's plants. But it remains a race against time.

Professor Peter Raven is one of the world's most influential and respected botanists. Based at the Missouri Botanical Garden in the Midwestern city of St Louis, he has been involved in promoting plant conservation throughout his career: 'My early days were spent botanizing in California – one of the world's richest areas of plant diversity. I remember the excitement of discovering new species of *Clarkia* and *Arctostaphylos* as a teenager. In the 1950s it was hard to imagine that the world would ever change much. By the mid 1960s, though, it had become obvious that urbanization and the loss of habitat was leading to the destruction of some of my favourite wild places and plants. Over the past 60 years I have seen the relentless pace of vegetation and species destruction worldwide. We need to wake up to the long-term consequences.'

Botanic gardens around the world have an important role to play in plant conservation. Around 2,500 botanic gardens collectively grow about one-third of all known flowering plants, including many that are naturally rare or threatened with extinction. This is a huge resource for safeguarding the world's flora, providing an insurance policy against the loss of species in the wild and a supply of material for species and habitat restoration. The horticultural skills in botanic gardens are also immensely valuable in managing plant conservation projects – very often wild plant populations need to be managed to ensure their survival. Botanic gardens also provide excellent venues for the display of endangered plants that most people will never see in the wild.

## NATURAL BOUNTY OF FORESTS

Over 350 million people live in forested areas and over a billion people rely heavily on forest resources for their livelihoods. Forests are important to local and national economies with about 60 million people being employed in the forestry and wood industries. In Africa alone, firewood and charcoal provide approximately 70 per cent of energy requirements. The export of timber and other forest products such nuts, fruit and gum generates nearly 10 per cent of the economic product of African countries.

Vietnamese villagers harvesting firewood. As in many other parts of the world wood is the main energy source for cooking and heating.

Over 2 billion people rely on traditional medicines harvested from the forests and a wide range of other non-timber forest products (NTFPs) both for subsistence use and for sale. The value of NTFPs in global trade is currently estimated at US$4.7 billion annually.

It is generally recognized that maintaining functional populations of wild species in their natural habitats is the primary goal of most species conservation. Thirty years ago, Dr David Bramwell, a leading plant conservationist and botanic garden director, wrote: 'We are, when talking about conservation in gardens, I hope talking of the conservation of representative gene pools for practical research and for rehabilitation in natural ecosystem reserves and not just as museum collections ... We appreciate that the ideal solution is conservation in the field, in natural habitats; but the worldwide problem of destruction of natural vegetation and extinction of plant species is such a serious one that we cannot wait for the utopian situation where all countries have adequate conservation legislation and the goodwill and economic means to enforce it.'

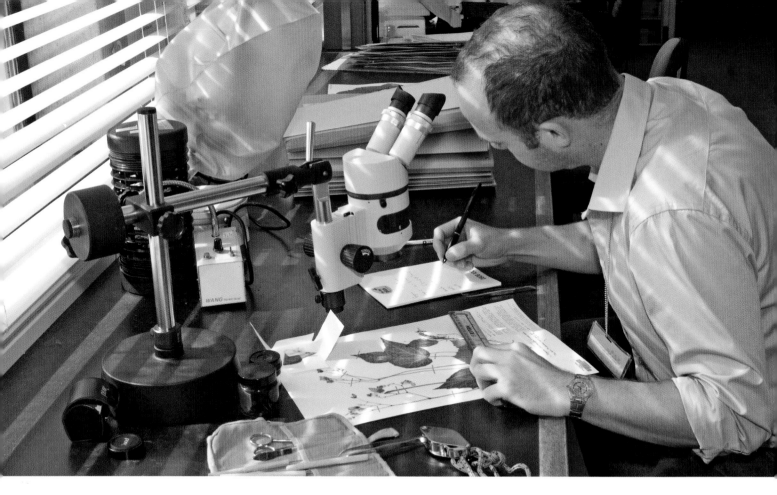

Inside the herbarium at the Royal Botanic Garden Edinburgh. Scientists working in herbaria around the world are cataloguing plant species as a basis for Floras, field guides and conservation assessments.

## SAVING SPECIES

There are many examples of plant species that have been rescued from extinction by botanic gardens. *Sophora toromiro* is one of the many species on a life-support system. This species is extinct in the wild and is now known only from a few valuable trees that survive in botanic garden collections. This small tree or shrub, commonly known as the toromiro tree, has pale green deeply divided leaves and yellow pea-like flowers. The last wild tree grew on the slopes of the Rano Kao crater at the south-west tip of Easter Island (Rapa Nui) one of the most isolated places on earth. When the tree was first discovered by western botanists, over two centuries ago, it formed thickets on the volcanic slopes. Reinhold Forster, travelling with Captain Cook on his second world voyage in 1772, wrote: 'By resting several times we were at last enabled to reach the summit of the hill, from whence we saw the sea to the west and the ship at anchor. The hill (Hanga Roa) was covered with a shrubbery of the mimosa [*S. toromiro*], which grew here to the height of eight or nine feet, and some of whose stems were about the thickness of a man's thigh.'

Toromiro was traditionally very important as the only native tree species and source of wood for the people of Easter Island. It was used for making canoes, house frames and the famous wooden carvings. The main reason for its extinction has been grazing of the seedlings and removal of bark by domestic sheep and cattle introduced to the island by European sailors and settlers. The remote island is now mainly covered in degraded grassland. The traditional Polynesian culture remains strong but there is no longer any native wood for use on the island. Thor Heyerdahl, the Norwegian archaeologist and explorer, is believed to be the last visitor to Easter Island to see *S. toromiro* in its natural habitat. Between 1955 and 1956 he collected seeds from the last remaining tree. It is from this collection that the present stocks of cultivated *S. toromiro* descend. Eighteen botanic gardens grow the species and there are hopes of reintroducing the tree to its native island habitat. Other island *Sophora* species that are threatened with extinction include Crusoe's mayu-monte (*Sophora fernandeziana*) and Selkirk's mayu-monte (*Sophora masafuerana*) both of which are confined to single islands in the Juan Fernandez group. Fortunately *Sophora* species are easily grown from seed and so there is hope for their long-term managed recovery.

Another famous species safe in cultivation but already extinct in the wild is the Franklin tree (*Franklinia alatahama*), an attractive member of the camellia family. This species was discovered in boggy areas along the Alatahama River in Georgia, USA, by John Bartram and his son William in 1765. The Bartrams were well known plantsmen from Philadelphia. The only species in its genus, *Franklinia alatahama* was named after Dr Benjamin Franklin, one of the drafters and signatories of the US Declaration of Independence. It is thought that the Franklin tree was naturally rare in the wild at the time of discovery and that over-collection by nurserymen contributed to its extinction. This beautiful species with large creamy white flowers that smell like orange-blossom was last seen in the wild 200 years ago.

Another species, *Deppea splendens*, sometimes known as cristobal, is a much more recent extinction in the wild. This attractive plant was first collected as herbarium specimens in 1972 as part of work to document the flora of Chiapas a state of Mexico. Seeds from the wild plants were collected in 1981 and given to botanic gardens in California. The species was not described by scientists until 1987. The previous year the only known site for this small tree, a canyon on the south slope of Cerro Mozotal in southern Chiapas was cleared for farmland and *Deppea splendens* is now presumed to be extinct in the wild. There still remains

propagation of threatened plants in botanic gardens is vitally important for their long-term conservation

a hope that it might be rediscovered on other nearby mountains in Chiapas or neighbouring parts of Guatemala but in the meantime this attractive plant with orange tubular flowers survives only in cultivation.

There are many different ways to help save plant species from extinction. Field research is needed to study the distribution and ecology of rare and threatened plants, and it is crucial that natural habitats are protected. For economically important plants harvested from the wild, sustainable-use strategies are required to prevent over-exploitation, and in some cases legislation is needed to prevent the removal of wild plants. For species over-harvested for international trade, international agreement to tackle the threat as provided by the Convention on International Trade in Endangered Species (CITES) is very important. CITES helps protect not only animals that are poached for international trade but also a range of plants including all orchids, cacti and cycads. Sometimes the causes of decline for wild plants need to be researched and for very rare species studies of the genetic makeup and reproductive strategies are important. Long-term storage of seeds or living plants in documented collections by botanic gardens is an important component of conserving the world's flora.

Propagation of threatened plants in botanic gardens is a vitally important component of their long-term conservation. When numbers of individuals have fallen perilously low in the wild horticulturists can step in to bulk up the numbers and enable the perpetuation of the species. Understanding the reproductive biology for such species is important. In some species male and female flowers are on different plants; in other cases flowers of both sexes are on the same plant and others have male and female parts on the same flower. The situation may be further complicated by mechanisms which prevent self-fertilization, and thus stop seeds from forming.

Propagation by seed is the best method to perpetuate genetic variability of a plant species but other forms of propagation have to be employed for some of the world's rarest plants. Micropropagation – growing plants from seed or small pieces of tissue under sterile conditions – has been immensely useful in plant conservation over the past thirty years or so. One of the first micropropagation units for plant conservation was established at the Royal Botanic Gardens, Kew in 1974 to propagate plants that are rare, endangered or difficult to grow conventionally. Over 3,000 different plant species have since been grown in Kew's Micropropagation Unit, including some of the world's very rarest.

## THE CHILEAN CROCUS

The fabled Chilean crocus (*Tecophilea cyanocrocus*) has beautiful gentian-blue, fragrant flowers. It was long thought to be extinct in its native alpine habitat as a result of grazing and over-collecting for horticulture. A comprehensive restoration project was initiated by the Royal Botanic Gardens, Kew, the UK Alpine Garden Society and the Chilean forestry authority. Using cultivated material Kew scientists worked to develop seedling populations with as much genetic diversity as possible. In 2001 the crocuses were found clinging to survival in the wild, growing amongst spiny shrubs in a field south of Chile's capital Santiago. The cultivated material provides an insurance policy to support these vulnerable populations.

## THE SPECIAL ROLE OF BOTANIC GARDENS

Botanic gardens are involved in many aspects of plant conservation but the scale of the task globally is immense. Perhaps the single most important thing that botanic gardens can do is to promote the importance of plant diversity and spread the message of plant conservation so that more people become involved in the task. Frequently, botanic gardens are situated in urban areas providing an opportunity for them to educate and inform a wide range of people who may otherwise have little contact with the natural world.

Singapore Botanic Gardens is one the world's best known botanic gardens. Located in the centre of the thriving city state this garden still conserves a

Children are encouraged to appreciate plants in Singapore Botanic Garden, providing an important experience of nature in their urban environment. Here they are fascinated by the touch-me-not plant, *Mimosa pudica*.

remnant of Singapore's rainforest within its grounds and has a highly important educational role. Established in 1859, the gardens were initially developed to cultivate useful plants reflecting the importance of spices to the local economy. Singapore Botanic Gardens' early reputation as a world centre for tropical botany and its economic application was established by the first director, Henry Nicholas Ridley. As well as promoting tropical agriculture, Ridley was a keen field botanist who travelled throughout the Malay Archipelago. He was very aware of the need to conserve tropical rainforests and their biodiversity alongside the development of rubber and palm oil, which he helped to establish as Asian plantation crops.

Part of the land on which the Singapore Botanic Gardens were developed had previously been cultivated and part was covered with pristine rainforest. The remnant patch of rainforest within the garden has 314 plant species surviving in the four-hectare forest patch. The Singapore Botanic Gardens provide an invaluable green haven close to the bustle of the city centre. Every year the gardens attract over two million visitors, more than half of which are from overseas. Along with the manicured lawns, magnificent orchid and ginger displays visitors can experience the feel of the rainforest and find out about the rich plant diversity of Southeast Asia. Currently, special educational programmes and talks are provided to 19,000 children and 700 adults a year and recently a Children's Garden was opened, 'to create happy memories in a fun and wonderful place and develop an appreciation for plants and the environment.'

Across the world another botanic garden with a very important educational role is the Brooklyn Botanic Garden in New York. This garden was established in 1910 from a reclaimed waste dump and now covers 52 acres in central Brooklyn. The garden's chief mission is to educate the public about plants and, as an extension, inform people about ecology and awareness of the environment. The Director of the garden, Scot Medbury, who has worked in botanic gardens all over the world, considers this to be immensely important. 'Some of the tree species growing in the garden are close to extinction in the wild, giving us the chance to introduce city-dwellers to real-world conservation issues. The genius of this place is that many people are inspired to transfer an initial enthusiasm for growing and understanding plants to becoming better stewards and advocates of the natural environment.' A Children's Garden programme has been in operation at Brooklyn Botanic Garden since 1914 – the first garden of its kind in the world. A more recent scheme, Project Green Reach, established in 1989, aims to encourage the study of botany and environmental science in low-income urban communities.

Botanic gardens collectively share resources and expertise through national, regional and global networking. In this way their conservation achievements are replicated and amplified. At a global level, botanic gardens work together through membership of and support for Botanic Gardens Conservation International (BGCI). This organization was established 30 years ago at first operating as part of IUCN. The initial aim was to find out where globally threatened plant species were secure in conservation collections – a role that has been refined and remains important today. BGCI has developed the PlantSearch database as a means to identify plants in cultivation in botanic gardens. The

The children's garden at Brooklyn Botanic Garden. The first botanic garden to set up a special garden for children, Brooklyn has a wide range of educational activities designed for children, adults, families and teachers.

database currently holds records for over 150,000 taxa, provided by nearly 700 botanic gardens. The plant records are linked to the IUCN Red List of threatened plant species so that botanic gardens can check which species they have in their collections are of global conservation concern. The database is also an important conservation planning tool enabling botanists to see which endangered plant species are not yet held in documented collections and also where material might be available for recovery programmes.

BGCI and its members have strengthened their commitment to plant conservation over the years and have played an important role in developing plant conservation policy that goes beyond the botanic garden community. Botanic gardens have, for example, been involved from the outset in the development of Global Strategy for Plant Conservation (GSPC) an international agreement under the auspices of the Convention on Biological Diversity (CBD). This convention is the leading international agreement for the conservation of species and ecosystems signed up to by nearly all the world's governments. The idea to develop a specific agreement for plant conservation came up at an international botanical congress held in St Louis, US in 1999 because plant conservation was considered to be lacking political and financial support. Botanic gardens together with international conservation organizations worked collaboratively to develop the Strategy and this was agreed by all the governments involved in the CBD in 2002.

## THE FUTURE

The ultimate goal of the GSPC is to halt the current and continuing loss of plant diversity. The Strategy consists of 16 targets to be met by 2010. Achieving the GSPC Targets will contribute to the overall 2010 Biodiversity Target ,which calls for a significant reduction in the rate of loss of biodiversity by this year. The GSPC targets cover *in situ* and *ex situ* conservation, sustainable production for agriculture and forestry, valuing traditional knowledge of wild plants and their uses, sharing skills and resources and informing the public about our global reliance on plants.

Implementation of the GSPC is the responsibility of national governments who have signed up to the CBD and agreed the Strategy. Botanic gardens have played a major role in supporting the development of national strategies for plant conservation and then implementing these. China's Plant Conservation Strategy has, for example, been developed by the leading botanic gardens within

**storage of as much wild plant diversity as possible in *ex situ* collections is an immediate imperative**

THE PLANT EXTINCTION CRISIS

the country working with different ministries of the Chinese government. BGCI helped to facilitate this process based on experiences in the UK and other countries around the world. As China has ten per cent of the world's flora, national progress towards meeting the GSPC targets will have significant global impact.

The progress towards meeting the targets of the GSPC has been impressive and the agreement has united different groups who care about plants with a common agenda but much more work needs to be done. Looking beyond 2010 the targets are being re-written to take into account the reality of climate change.

As David Bramwell points out, 'The plant conservation agenda needs to be re-thought. With rising temperatures and changing weather patterns on an unprecedented scale even currently common species are likely to come under threat. Protected areas may no longer be in the best place to conserve diversity if species shift their ranges. Storage of as much wild plant diversity as possible in *ex situ* collections is an immediate imperative. We cannot give up on the wild – but do we know what or where the wild might be?'

The successful implementation of the GSPC will help to ensure the security of the world's vegetation and thereby reduce carbon emissions more than equivalent to those generated by the world's combined transport systems. Maintaining ecosystems as carbon sinks and as reservoirs of genetic and species resources for the future, remains vitally important. The role of botanic gardens as modern-day arks supports and complements ecosystem conservation by helping to maintain the species that are the building blocks of ecosystems and the key to our future.

Increasingly sophisticated techniques are being developed to store seeds and genetic material from rare plants so that cultivation and restoration of wild populations remain options for the future.

The Royal Botanic Gardens, Kew has long played a pivotal role in the development of botanical knowledge and of plant conservation worldwide. The current Director, Professor Stephen Hopper, considers conserving and restoring the world's imperilled flora to be Kew's most important role. 'Everything we do in exploration, research, education, horticulture and seed banking must be geared towards saving plant diversity. The challenge is enormous in the face of global climate change, but conserving and restoring the wealth of plants for human benefit is the true remit for Kew and for many other botanic gardens around the world.'

**Previous page:** The lady's slipper orchid (*Cypripedium calceolus*) has become an important symbol for plant conservation. Its beauty has nearly led to its downfall in many countries but the Royal Botanic Gardens, Kew, is helping to ensure this species survives.

Kew's strong focus on plant conservation developed in the 1970s. During that decade of growing environmental awareness, Kew began the process of documenting the conservation status of plants worldwide on behalf of IUCN. The first international Red Data Book for plants was published in 1979 and contained 250 examples of globally threatened species. The collection of information on threatened plant species fed into international policy and Kew helped to draw up the first list of plant species to be protected by CITES. During the same period, two conferences on practical plant conservation were organized by Kew. These brought together experts from around the world and called for coordinated action.

As one response to this the Botanic Gardens Conservation Coordinating Body was established in 1979 as a Specialist Group of IUCN with a small secretariat at Kew. Later the secretariat evolved into the independent organization, BGCI. Also in the 1970s Kew's Micropropagation Unit was established – one of the first facilities to propagate the very rarest plant species.

Kew's role in plant conservation today encompasses botanical exploring and desciribing plants in remote parts of the world; documenting the status of rare

and threatened plants and their habitats; researching ecologically and economically important plants; supporting and informing conservation policy; conserving plants *ex situ* through living collections, seed banking and DNA banking; and developing innovative conservation solutions. Much of this goes on behind the scenes but Kew's celebration of plant diversity and promotion of the historical significance of plants and gardening is something that all visitors to the 250-year-old garden can enjoy.

The alpine house at Kew has stunning displays of plants from mountain areas around the world. Many alpine plants are naturally rare and are increasingly threatened as climate change affects their habitats and lifecycles. Cultivating alpines increases understanding of their growth requirements.

## KEW'S ORIGINS AND LEGACY

In 1759 Princess Augusta, the mother of King George III, hired William Aiton to develop a physic garden as part of her Kew property close to the River Thames. This garden of around 4 hectares was to form the nucleus for the development of the Royal Botanic Gardens, Kew. It was not until after Princess Augusta's death in 1772 that Kew's international reputation was established by Sir Joseph Banks. An enthusiastic wealthy plant collector and botanist, Banks went on several expeditions as self-funded naturalist, including James Cook's circumnavigation of the globe on the *Endeavour* (1768–71). He served as

Herbarium specimens are a major resource for checking plant names. They also provide a wealth of data on the distribution of species that can be used in conservation planning. This specimen is a type specimen - the original material used in naming a species.

honorary and unofficial director at Kew from 1773 until his death in 1820. During that time, he oversaw collections from around the world and Kew became known for plant collecting on an international scale.

A period of apathy followed the death of George III, but in 1841 Sir William Hooker was appointed as the first official director. He was succeeded by his son, Sir Joseph Hooker, in 1865. Under the Hookers' leadership, Kew was reinvigorated and became a centre for scientific research and the exchange of plant specimens. The Herbarium, established in 1853, now houses around 7 million specimens and is of enormous global signficance.

Joseph Hooker was a close friend and supporter of Darwin, who bequeathed money to Kew for the development of *Index Kewensis*, a list of all flowering plant names recorded in botanical literature from the time of Linnaeus. This list is still added to today, and is now accessible online. Every time a new species is described or a species is renamed as knowledge about its taxonomic relationships is enhanced, the name is added to *Index Kewensis*. The Herbarium at Kew also provides the largest reference collection of dried plant specimens in the world, a vital resource which is the basis for understanding the world's plant diversity and cataloguing all plant species. Professor David Mabberley is the current Keeper of the Herbarium, maintaining the venerable tradition established in 1853.

Mabberley has long been fascinated by the naming of plants and sees this as a crucial first step in plant conservation: 'The Global Strategy for Plant Conservation called for a complete working list of all plant species by 2010. Kew has worked hard with other botanic gardens, universities and individuals to achieve this, building on our legacy of exploring and documenting plant diversity since the time of Banks. We now have Floras and plant checklists for most parts of the world. Harmonizing these and getting the names right provides a baseline for conserving and monitoring the world's flora.'

## ORCHIDS: THREATENED BY THEIR OWN BEAUTY?

Orchids are one of the iconic groups of plants that are being studied and cared for by Kew. The natural diversity of orchids is staggering: an estimated 18–30,000 species grow worldwide in virtually every region of the globe. It is probable that up to one third of all tropical orchids and many temperate species from regions such as Australasia have yet to be fully catalogued scientifically. With stunning flowers and an exotic allure, many attractive species of orchid are threatened with extinction as a result of habitat loss and the voracious attention of collectors.

The herbarium specimens and extensive living collections of orchids maintained by Kew are critical for improved understanding of the orchids of tropical regions. Kew's long-standing commitments to the preparation of the Floras of tropical Africa are to be completed in the next few years. The attention of Kew's specialists has turned also to tropical Asia, which has a very rich diversity of orchids. In collaboration with the National Herbarium, Netherlands, and institutes in Southeast Asia, Kew is helping to describe and record the orchids of Malaysia and Indonesia. This published information helps to form the basis for conservation policies and programmes within these countries. Kew also carries out practical conservation action for orchids within its own research laboratories and grounds.

One of Kew's leading orchid experts is Dr Mike Fay, head of its Genetics Section. Fay has had a life-long interest in orchids and is Chair of the IUCN/SSC Orchid Specialist Group as well as running his laboratory work. The Orchid Specialist Group consists of over 200 experts worldwide who care about the

**many species of orchid are threatened with extinction as a result of habitat loss and the voracious attention of collectors**

conservation of orchids and act collectively to save species from extinction, either through advice or practical action.

There are around 50 wild orchid species in the UK. Approximately a third of these are thought to be threatened and 10 are protected by national conservation legislation. Several species have populations of less than 100 individuals, and changes in land use continue to take their toll. Theft from wild populations is also a problem. Fay and other Kew botanists have been working to save the very rarest UK orchids since 1983 when the Sainsbury Orchid Conservation Project began at the Micropropagation Unit. The intention throughout this project has been to use the latest research and conservation techniques to reverse the trend of extinction for British orchids.

The lady's slipper orchid (*Cypripedium calceolus*) is the most visually stunning and rarest of all the orchids native to the UK. Although widespread on a global scale the species is threatened with extinction in many parts of its range. In the UK, the lady's slipper orchid once grew on grazed limestone grassland in Derbyshire, Yorkshire, Durham and Cumbria. Uprooting by gardeners, picking and trampling by botanists, and habitat modification due to increased grazing pressure led to a severe decline until only one flowering individual remained clinging on to survival in the wild. The flowers of this single plant were hand-pollinated for a number of years to encourange the production of seeds. Kew, working collaboratively with the UK's national conservation agencies, was given permission to collect small amounts of seed for micropropagation purposes. The first six seedlings were planted back into the wild in 1987. One of these flowered for the first time 13 years later. Now over 1,500 seedlings have been planted at various locations. Further plants, derived from wild stock, exist in cultivation at Kew and other carefully selected gardens.

Since 1992, a coordinated recovery programme has been in place for the lady's slipper orchid in the UK. This has included carefully monitoring and guarding the site of the remnant wild population and establishing material from the wild plant in cultivation. The genetic diversity of the cultivated material is taken into account when assessing the viability of seedlings selected for re-stocking the wild. Seed is maintained at Kew's Millennium Seed Bank at Wakehurst Place, and scientists continue to research the most appropriate methods to store seed for its long-term viability. Mature seed can be stored in seed banks but is difficult to germinate whereas immature seed that germinates well does not respond well to storage.

The Micropropagation Unit at Kew pioneered new techniques in the battle to save species from extinction. Micropropagation, growing plants from seed or small sections of plant material under carefully controlled sterile conditions, means that very rare species can be propagated in bulk. Kew has grown over 3,000 plant species in this way.

# PAPHIOPEDILUM

The tropical slipper orchids in the genus *Paphiopedilum* are one group that have been particularly targeted by collectors. Over 60 species of *Paphiopedilum* grow in eastern Asia from India to New Guinea and the Solomon Islands in the Pacific. Many species are naturally rare, growing, for example, on isolated limestone outcrops in rainforest areas. The fabled *Paphiopedilum rothschildianum*, perhaps the most spectacular of all tropical slipper orchids, is only known from Mount Kinabalu National Park in Sabah, Borneo.

Orchids are subject to the whims of fashion. The stunning flowers of *Paphiopedilum* orchids have attracted the attention of orchid collectors since Victorian times. In the 1960s the discovery in Thailand of *P. sukhakulii*, a new species with an unusual flower shape, led to renewed interest in the genus and fuelled an international demand for wild-collected plants. This was repeated twenty years later when the bright yellow-flowered *P. armeniacum* and huge pink-flowered *P. micranthum* were discovered in Yunnan, China. Subsequently collectors targeted the endangered slipper orchids of Vietnam, of which there are over a dozen species. These orchids are directly threatened by illegal international trade despite CITES protection. Kew has published an account of the genus *Paphiopedilum*, publicizing its conservation needs, and is actively supporting its conservation through field work and training programmes.

Four other endangered British species, the military orchid (*Orchis militaris*), monkey orchid (*O. simia*), lizard orchid (*Himantoglossum hircinum*), and fen orchid (*Liparis loeselii*), have been successfully reintroduced into selected sites as have more common species of *Anacamptis* and *Dactylorhiza*. In each case Kew scientists carefully grew plants from the microscopic seeds and helped to prepare them for reintroduction. The orchid propagation techniques developed by Kew have been used successfully elsewhere in the world to raise seedlings for conservation and horticultural purposes.

Increasingly, understanding the genetic make-up of plants can be crucial to determining the best way to conserve them. Scientists working on conservation genetics in Kew's Jodrell Laboratory use a range of genetic fingerprinting techniques to investigate population genetics of orchids, and the data collected are being used to inform conservation programmes for endangered species.

## KEW'S LIVING COLLECTIONS

Kew grows around 30,000 different species in its documented living collections. Keeping careful records of the plants grown in the glasshouses and grounds is particularly important for plants of conservation concern. 'Some of the plants we are able to grow are precious remnants of formerly more abundant species,' says Kew's Curator, Dr Nigel Taylor. 'We need to maintain records of the geographical provenance of our wild-collected plants if they are to be considered part of a scientific research and reference collection and of value for species conservation. Over the years Kew has helped reintroduce many different rare and challenging species including orchids, palms and trees such as the St Helena ebony (*Trochetiopsis ebenus*), which was reduced to two individuals on the remote South Atlantic island.'

A particular interest of Taylor's is the taxonomy and cultivation of cacti. He grew cacti as a child and was determined to develop a career in horticulture. The second Kew conservation conference in 1978, the year after Taylor joined Kew as a researcher in the Herbarium, gave him a detailed insight into plant conservation challenges worldwide and how botanic gardens can help in many aspects of conservation. Subsequently Taylor has helped document the conservation status of cacti in their natural habitats.

'Practical application of horticultural techniques is vital if we are to care for our threatened plant species,' he says. 'I have seen cacti on the brink of extinction in the wild whilst working in Mexico and Brazil. Around a third of all

The Princess of Wales Conservatory at Kew has displays of plants from 10 climate zones. The displays include species that have become very rare in the wild, including succulents from desert areas.

cacti are threatened with extinction and so botanic gardens need to work together internationally to ensure that such species are carefully cultivated, propagated and wherever possible returned to their native habitats.'

## THE MILLENNIUM SEED BANK

In addition to its extensive living plant collections, Kew also manages the world's largest collections of stored seeds – the Millennium Seed Bank. Inaugurated in 2000, Kew's seed bank now contains collections for around 97 per cent of the UK's native flowering plants, conifers and ferns. For orchids, which are usually reliant on symbiotic fungi for successful germination, Kew scientists working with collaborators in Kenya and Australia have pioneered a system of simultaneously cryopreserving (freezing) orchid seeds with their fungal partners. The Millennium Seed Bank is also helping to coordinate a network of European seed banks, which seeks to share expertise and facilities, coordinate the setting of priorities and therefore avoid duplication of effort across continental Europe. Further afield, the Millennium Seed Bank successfully set out to collect and conserve 10 per cent of the world's flowering plants and conifers (some 28,000 species), mainly from the arid areas of the world, by 2010. The target is now to bank 75,000 plant species by 2020.

Drylands cover a third of the Earth's land surface, including many of the world's poorest countries, and support almost one fifth of its population. The most immediate threat to dryland areas and the plant species that can survive these harsh conditions is desertification due to overgrazing and agriculture. The seeds stored by Kew will help to ensure the potential restoration of dryland areas in the future. Already Kew is helping to restore species in arid areas of Namibia, Kenya and South Africa.

## INTERNATIONAL COLLABORATION

Kew's seed banking activities around the world are undertaken collaboratively, ensuring that research, training and capacity-building relationships are developed with the countries where the seeds are collected. International collaborations are based on principles of the Convention on Biological Diversity, respecting national sovereignty and supporting national conservation strategies. Benefit-sharing, in the form of duplicate seed storage, data exchange, technology transfer and training are all essential components of the Millennium Seed Bank's work.

In addition to the seed banking partners in 54 countries around the world, Kew collaborates with local botanists in over 100 countries to support plant conservation. Strong links have been developed in some of the 'megadiverse countries' (those with an especially rich diversity of life), such as Brazil and China, where botanists are working together to slow the relentless pace of plant loss.

Chris Leon, one of Kew's ethnobotanists, is working with Chinese colleagues to develop a comprehensive reference collection of Chinese medicinal plants. This ambitious undertaking involves cataloguing and photographing over 2,000 plant species that are used medicinally in China. The aim of this project is to help conserve the species in their natural habitats and increase knowledge of their use. Equally important is work to improve the safety of traditional medicines. 'There is very great interest in how China's traditional medicines

botanic gardens need to work together to ensure that species are carefully cultivated and propagated, and returned to their native habitats whenever possible

might help in healthcare in the West,' says Leon. 'With over two thousand years of experience in treating ailments, Chinese medical tradition has developed in a different way to our practices in Europe. There is so much to learn for our mutual benefit. We need to understand which species are used, how they are processed and how we can ensure a continuing supply. In total some 11,000 Chinese plants have some medicinal use of which 600 are particularly important.' One specific challenge is to conserve the wild species at locations where they are renowned for being of especially good quality. So called *didao* populations of medicinal plants are highly prized as effective medicines, which places a particular demand on their wild populations.

Chris has travelled extensively in China, working in 19 of its 22 provinces to collect and photograph medicinal plants and see at first hand how the various species are used. Different species are more popular in different parts of the country and about a third of the ethnic minority groups in China have their own traditional medicines. There is, nevertheless, a standard *Pharmacopoeia of the People's Republic of China*, which documents the wealth of knowledge for the country as a whole. Among the very rare and valuable plants Chris has photographed in the wild are the fritillaries *Fritillaria cirrhosa* and *F. delavayii*, orchids *Bletilla striata* and *Dendrobium nobile*, and *Aquilaria sinensis*, a source of agarwood.

Back in the UK, Chris runs the unique Traditional Chinese Medicine Authentication and Conservation Centre. The centre's work involves identifying plants entering the UK for use in traditional Chinese medicine, to promote patient safety and detect substitute herbs that may be less effective or even harmful. Over 3,000 clinics in the UK now offer Chinese traditional medicines for a range of ailments from eczema and psoriasis to depression and insomnia. Typically, the treatments involve a complex mixture of plant ingredients, many of which can be tricky to identify. Using the reference herbs collected by the centre, Kew's researchers are developing new authentication methods, using chemical and molecular fingerprinting techniques, to identify even the most challenging of ingredients, be they in the form of dried leaf fragments or tableted extracts in patent remedies. These same techniques are also being used to detect the use of rare and protected plant species, thereby helping with the enforcement of CITES and other conservation legislation.

## LOOKING TO THE FUTURE

The wild plants that Kew is helping to conserve – whether orchids in the UK, plants vital for healthcare in China or the dryland and island species – are coming under increasing threat with global climate change. In 2003, Kew was awarded UNESCO World Heritage Site status in recognition of its leading contribution to the study of plant diversity and economic botany around the world as well as its fine landscape gardens. Kew's World Heritage status recognizes that the landscape gardens and the edifices created by celebrated artists reflect the beginning of movements which were to have international influence. Although the grounds continue to be an inspiring location for the celebration of plants and gardening but the international focus and influence of Kew's work is now centred on conserving and restoring wild plants. The ambitious Breathing Planet programme launched in 2009 sets out a 10 year plan for Kew's work worldwide.

The Royal Botanic Garden Edinburgh dates back to 1670 when two Edinburgh doctors, Andrew Balfour and Robert Sibbald, leased a small enclosure near Holyrood Abbey to establish a medicinal plants collection. The purpose of the garden was to teach botany to students, train apothecaries, produce a Scottish pharmacopeia and cultivate foreign plants. The garden moved site three times during its history and was established at its present site in Inverleith in 1820–23. A very wide range of temperate and tropical plants can be seen at Inverleith and the Edinburgh collection is supplemented by the gardens at Benmore, Dawyck and Logan. The Royal Botanic Garden Edinburgh retains its interest in medicinal plants but its research and conservation interests have expanded enormously. One group of particular interest is conifers.

**Previous page:** The palm houses at Royal Botanic Garden Edinburgh were built in the Victorian era and are still popular with visitors today. The glasshouse in the foreground dates back to the 1960s.

Conifers are formidable plants. More ancient than flowering plants, conifers have been on Earth since the Triassic period and their ancestors were widespread in the late Carboniferous period. They have a particular importance in the global ecosystem because they can thrive at high latitudes where the growing season is too short for broadleaved trees. The oldest and tallest trees in existence are all conifers and sadly they are now also some of the rarest: approximately one third of the 630 conifer species are threatened with extinction in the wild. We know this because there is a very active group of conifer experts involved in conservation planning. At the forefront of action for conifers is the Royal Botanic Garden Edinburgh.

The International Conifer Conservation Programme was established in Edinburgh in 1991. From the outset the programme was designed to integrate different approaches to conifer conservation to ensure the best chance of saving species from extinction. Taxonomic studies, field surveys, habitat protection, *ex situ* conservation, training, and restoration of wild populations are all important parts of the process. Martin Gardner runs the programme and has worked on conifer conservation throughout his time at Edinburgh. He believes

Looking out across 'The Botanics' in Edinburgh. The botanic garden is a much-loved asset of the city.

conifers are often dismissed as Christmas trees or leylandii hedges but they are fascinating trees of immense ecological and economic value. 'Some will be difficult to conserve because we have a very limited genetic base to work with,' he says, 'but given the will there is no need for any conifer to become extinct.'

## SAVING THE ALERCE

Gardner has travelled the world researching conifer species in the wild – an essential component of any conservation programme – with Chile and New Caledonia the focus for much collaborative work. The impressive temperate rainforests of Chile have nine conifer species, some of which are severely reduced in the wild. Alerce (*Fitzroya cupressoides*) is one such tree.

Named after Captain FitzRoy of HMS *Beagle*, the ship on which Darwin sailed, individuals of this extremely slow-growing and massive conifer can live for over 3,000 years. Alerce's valuable timber has contributed to the downfall of the species. It is now considered to be Endangered and is listed on Appendix I of CITES in an attempt to ban international trade in wild-harvested timber.

Alerce was introduced to the UK in 1849 by William Lobb and although relatively uncommon in cultivation it can be seen in some botanic gardens, arboreta and country estates. The International Conifer Conservation Programme, working with the University of Edinburgh, discovered that 99 per cent of alerce trees in cultivation in Britain are derived from Lobb's original introduction – effectively all cuttings from a single female plant. Because genetic diversity is essential to ensure the long-term viability of the species the alerce specimens in the UK are of limited conservation value. Scientists have been collecting seeds and cuttings from across the natural range of alerce to broaden the genetic base of plants in cultivation.

Cones of alerce, a magnificent conifer of Argentina and Chile that is Endangered as a result of logging and destruction of its forest habitats

Once seeds or cuttings of any conifer species have been collected from the wild with full permission and local collaboration, the plants are propagated at the nursery at the Royal Botanic Garden Edinburgh. Careful notes are taken on methods of propagation and rate of growth of the young plants to ensure that this information is available for future conservation projects. Once the material is ready for distribution, plants are sent to a network of 'safe sites' across the UK. Alerce seedlings and cuttings have been distributed to 44 sites.

A particularly ambitious planting of alerce can be seen at one of Royal Botanic Garden Edinburgh's satellite gardens, the Benmore Botanic Garden. This splendid 50-hectare mountainside garden in Argyllshire with a warm, wet climate, shelters some of the tallest trees in Britain. Native lichens and mosses clothe the branches of exotic conifers that have been planted over the past 150 years. The avenue of giant redwood (*Sequoiadendron giganteum*), a species that is considered Vulnerable in the wild, is an impressive sight as visitors first enter the garden. The avenue was planted in 1863 by Piers Patrick, a wealthy American who owned the Benmore estate at the time. The next owner was James Duncan who made his money from refining sugar and planted over six million trees at Benmore. He was followed by the Younger family of Edinburgh brewers, who continued the tree planting tradition and whose descendants retain a link with Benmore and its conservation work today.

Within Chile, alerce is found in the Andes and the Coastal Range. Ten years ago small forest fragments with young *Fitzroya* populations were found in the area of Puerto Montt, in the Central Depression of Chile between the two mountain ranges. Shortly after their discovery botanists from the Royal Botanic Garden Edinburgh, Edinburgh University and the Universidad Austral de Chile

worked together to help restore these forest fragments, which were heavily threatened by urban sprawl. Seed collected from the local alerce trees was raised in nurseries and then planted out to supplement the wild populations. Constant vigilance will be needed to nurture the young trees and enable the decline of the species to be reversed.

Giant redwoods, planted in 1863, form a majestic avenue at the Benmore Botanic Garden in western Scotland. In the wild this magnificent species is now considered to be Vulnerable. Benmore grows a range of threatened conifer species that thrive in this area of high rainfall.

## OTHER CHILEAN CONIFERS

Cipres (*Pilgerodendron uviferum*) is another of the South American temperate rainforest conifers that grows in Chile and Argentina and is now in trouble in the wild. Wood from this species is valued for its durability and resistance to decay. It has been used in all types of building construction and to make bridges, boats and furniture. Large-scale destruction of Chile's temperate forest over centuries and exploitation of this species in particular has led to its current Endangered status. Illegal harvesting still takes place in many forests. Burning

Monkey puzzle trees in the wild. Commonly grown in suburban gardens, *Araucaria araucana* is native to Argentina and Chile, where it is valued for its timber and edible seeds.

and grazing have prevented regeneration, further contributing to the species' decline and cipres has been placed on Appendix I of CITES, which has reduced international trade. Now protection and restoration of its natural habitats is urgently needed. Within Chile, the Universidad Austral de Chile and Corporacion Nacional Forestal has developed demonstration plots for the sustainable management and restoration of cipres forest working with local forest owners. Edinburgh is helping with initiatives to support the conservation of *Pilgerodendron uviferum* and this is one of the species now being grown at Benmore, in patterns that mimic the conditions in the wild.

The monkey puzzle tree (*Araucaria araucana*) is another of the conifers grown in the Chilean rainforest planting at Benmore. Commonly associated with suburban gardens, the monkey puzzle is less threatened in the wild than alerce and cipres but nevertheless it is in trouble in its natural habitats. Restricted to parts of Chile and Argentina, it has declined as a result of logging, fire and other pressures on its natural environment and is now considered Vulnerable by IUCN.

*Araucaria araucana* has great historical and social importance in its native countries. The seeds form an important part of the Mapuche indigenous people's diet in Argentina and Chile. One group of Mapuche people, the Pehuenche, derive their name from 'pehuén', the local name for the monkey puzzle. Between November and December female flowers start growing as spherical green cones. Each cone releases between 120–200 seeds called 'piñones' which are harvested between February and May. During the winter months the seeds provide the main carbohydrate source for the Pehuenche people. Martin Gardner says that, 'the seeds collected from the wild are not only vital for broadening the genetic base of monkey puzzles in cultivation but are also delicious when eaten. They taste like floury hazel nuts.'

## INTERNATIONAL COLLABORATION

Botanists from Edinburgh have carried out extensive fieldwork in central and southern Chile, working collaboratively with local institutions. Work has concentrated mainly on 46 very rare and threatened species in this global biodiversity hotspot, including 7 of the native conifers. Detailed information has been collected on the distribution and conservation requirements of each species and strategies for conservation have been developed working

# BENMORE'S CHILEAN RAINFOREST

*Eucryphia cordifolia* is native to Chile and Argentina, where its natural habitats are under threat from logging and deforestation. It grows well as a garden plant and is available from commercial nurseries.

Chilean conifers have been planted on a five-hectare site on the steep sloping Creachan Mor, with each species planted along the slope in sequence to reflect their altitudinal zone in the wild. Alongside the conifers other rarely cultivated and threatened Chilean trees are being grown, such as *Nothofagus alesandrii*, the magnificent white-flowered *Eucryphia cordifolia*, and shrubs such as *Orites myrtoidea* and the spectacular *Berberis trigona*.

Another attractive species, *Embothrium coccineum*, was collected as seed on the Chilean island of Chiloé and flowered successfully at Benmore after five years. The flame-red flowers of this fire-bush add colour to the developing temperate rainforest vegetation. Benmore's curator, Peter Baxter, has been involved in Edinburgh's fieldwork in Chile and so has first-hand experience of the growing conditions in Chile's rainforest. 'The rainfall feels pretty similar in southern Chile and Argyllshire,' he says. 'We can't match the natural species diversity of Chile's ancient forests but we aim to give our visitors a good impression of the species assemblages that grow there.'

The ornamental *Embothrium coccineum* grows wild in Chile (below right). It is grown at Benmore (top) to help recreate the feel of a Chilean temperate rainforest.

with local landowners. Simple propagation techniques have been developed at the Universidad Austral de Chile which can be used in restoration programmes. For *Araucaria araucana* a group of local organizations has been established to care for the species in an area called Villa las Araucarias. The local people are creating a small community reserve for restoration of the species. Elsewhere in Chile, the Royal Botanic Garden Edinburgh is supporting plans to re-connect fragments of *Araucaria* forest. Working with a small consortium of staff from the University of Valdivia and the London-based NGO, Rainforest Concern, the International Conifer Conservation Programme has helped to purchase 110 hectares of *Araucaria* forest in the Chilean Andes. This protected area, known as Nasampulli, is strategically situated between three national parks that contain large populations of *Araucaria araucana*. Plans are being made to establish a corridor between Nasampulli and these other protected areas.

## NEW CALEDONIA

Trees in the genus *Araucaria* are among the oldest conifer species, dating back around 70 million years to a time when they enjoyed a widespread distribution. Now they are all restricted to the southern hemisphere with 13 of the 19 species found on the Pacific island of New Caledonia – a botanical hotspot and another area of work for the Royal Botanic Garden, Edinburgh. Despite its small size, New Caledonia has 43 endemic conifer species, which makes it an attractive destination for a conifer enthusiast like Martin Gardner. He first visited the island in 1999 in order to establish research links and to see first-hand some of the extraordinary, if not bizarre, conifer species that New Caledonia is famous for.

Sadly, the high level of biodiversity on New Caledonia coincides with catastrophic environmental degradation, largely due to nickel mining and human-set fires. Many species, such as *Araucaria rulei*, are indicators of nickel-rich soils – they thrive in the very areas where mining takes place. *Araucaria rulei* is Endangered and is not currently protected in any nature reserves or other areas. Molecular techniques are being used to investigate how and why New Caledonia's high levels of conifer diversity evolved. Molecular research and taxonomy are also helping to guide conservation policy and provide vital baseline data for the International Conifer Conservation Programmes's publications on threatened conifers.

## UK PROJECTS: SCOTS PINE

Back in the UK, Scots pine (*Pinus sylvestris*) is one of only three native conifers. The wild populations in Scotland are the only truly native plants of this species in Britain. Scots pine is actually extinct in the wild in England and has been replanted. Although Scots pine is the most widely distributed conifer in the world, Scots pine woodland in the highlands of Scotland is now very patchily distributed and in need of conservation attention. Only about one per cent of the extent of the original woodland now remains. This special habitat includes rare species of plants such as twinflower (*Linnaea borealis*), one-flowered wintergreen (*Moneses uniflora*), and other wintergreens (*Pyrola media, Pyrola minor* and *Orthilia secunda*). Orchids such as creeping ladies tresses (*Goodyera repens*) and lesser twayblade (*Listera cordata*) are also found on the woodland floor. Scots pine woodland also supports a rich diversity of animal life. Wood ants (*Formica aquilonia*) form mounds of pine needles and other plant material up to a metre in height on the forest floor. Each social colony consists of up to half a million individuals. Birds of the Scots pine woodlands include the golden eagle (*Aquila chrysaetos*), black grouse (*Tetrao tetrix*), capercaillie, (*Tetrao urogallus*) and the UK's only endemic bird, the Scottish crossbill (*Loxia scotica*). Scottish crossbills feed on the seeds of the Scots pine as do red squirrels, mice and voles.

An expansion of roe deer and red deer populations Scottish Highlands over the past 150 years has been a major problem for the native pine woodlands. Deer browse on Scots pine seedlings and young trees, preventing the regeneration of the woodlands. Red deer also damage or kill young Scots pines by removing bark from trees with their antlers, particularly in late spring when the new season's antlers are shedding their velvet. No young Scots pine trees were able to survive in most parts of the Scottish Highlands until habitat management by fencing or intensive deer-culling measures was put in place over the last 20 years.

Because fragmentation of plant populations can cause major problems for the survival of species, scientists from the Royal Botanic Garden Edinburgh are looking at the reproductive ecology and genetics of isolated populations of plants within the Scots pine woodlands. The results will help show how plants respond and adapt to changing habitat conditions and inform plans for habitat restoration.

Scots pine growing in its native habitat in the Cairngorms National Park. A large part of this protected area is covered in native pinewoods.

# THE HIMALAYAN YEW

Linking Edinburgh's work on conifers with its interest in Chinese medicinal plants is an important project on the Himalayan yew (*Taxus wallichiana*). This tree is very important as a source of paclitaxel (Taxol), a cancer-inhibiting compound found in the bark of this and several other yew trees. As so often happens the unsustainable use of a plant resource has lead to the downfall of the species. *Taxus wallichiana* is threatened in the wild and is now listed on Appendix II of CITES in an attempt to prevent unsustainable levels of international trade. The species has three different varieties. Edinburgh is working with botanists in China to investigate whether the differences between the varieties are relevant to taxol production and to better understand their distribution for conservation planning.

### DEVELOPING CONSERVATION SOLUTIONS

Edinburgh's work on plant conservation in Scotland and overseas involves a range of approaches, including understanding the genetics and taxonomy of the species, working with partners to identify conservation needs and developing integrated conservation solutions with both *in situ* and *ex situ* techniques. Conifers are just one group of plants that benefit from this approach. Regius Keeper, Professor Stephen Blackmore stresses that: 'No one organization can conserve plants on its own; programmes like the International Conifer Conservation Programme and global networks like BGCI are essential if we are to succeed. And succeed we must since ours is a forest world dependent on photosynthesis – it cannot sustain us without the diversity of plants adapted to different situations around the planet'.

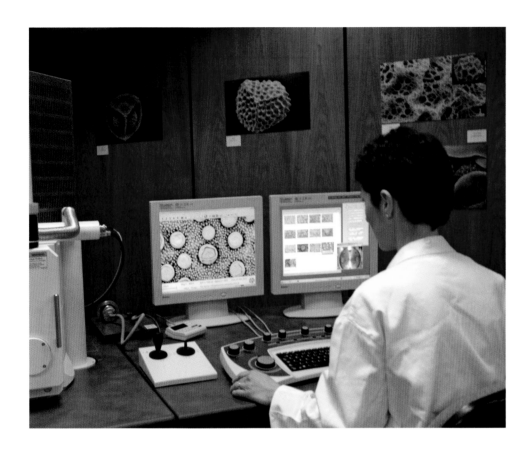

Botanists at the Royal Botanic Garden Edinburgh use a wide range of techniques to investigate and identify plants. The scanning electron microscope enables plant structures to be studied in minute detail and can help taxonomists distinguish between species when characters cannot be seen with the naked eye.

Plant conservation in France is organized in a very specific way to ensure an integrated national approach to the protection of native plant diversity. Concern about the threats to wild plants in the 1970s led to a carefully coordinated study of plants in need of protection and in 1975 the first Conservatoire Botanique was created at Brest in Brittany. Jean-Yves Lesouef was the driving force behind this new form of plant conservation centre dedicated to looking after rare and endangered plant species and their natural habitats. A unique federation of eleven Conservatoires Botaniques Nationaux now covers mainland France and beyond. One of the gardens in the network is in the plant-rich tropical island of Réunion in the Indian Ocean, another looks after the plants of the Caribbean islands of Guadeloupe and Martinique. Further Conservatoires are planned for New Caledonia and French Guiana.

**Previous page:** Subtropical plant displays in a glasshouse at the Conservatoire Botanique in Brest. This garden is working to conserve some of the world's most endangered and elusive plants, many of which grow on tropical, subtropical or Mediterranean islands.

Most of the Conservatoires concentrate on *in situ* protection of the threatened flora within their region of France, collectively taking care of the 486 priority species included in the French Red Data Book. The Conservatoire Botanique at Brest in Brittany is rather different in that the garden where it is based also acts as an ark for globally threatened plants from around the world. It probably grows more species that are extinct in the wild than any other botanic garden. The garden is a member of the Conservatoires Botaniques Nationaux and the botanic garden network Jardins Botaniques de France et des Pays Francophones.

Covering 47 hectares in a small valley of the Stang-Alar River, the garden has displays of endangered plants from all around the world acclimatized in a woodland setting. Beyond the garden walls, botanists take responsibility for plants in the wild within the biogeographical region of the Massif Armoricain, which covers Brittany, Pays de Loire and Normandy. Working with a network of 250 volunteers, the conservatoire coordinates the work of identifying, mapping and monitoring of the flora in this region of north-west France. In 2007, Brest finalized Red Lists for the Côtes d'Armor and Pays de Loire regions, providing a very valuable resource for conservation planning.

Over 200 species of special concern are monitored particularly closely in the wild. The conservatoire concentrates on *in situ* conservation action, working with several partners to protect plants in their natural habitats and supplementing them with carefully cultivated plants where necessary. Plants cultivated at the garden in Brest back up efforts to conserve plants threatened in their natural habitats and allow scientists to study the reproductive biology of rare and threatened species.

One species that has been rescued from extinction is an attractive white-flowered daffodil endemic to the Glénan islands, a small archipelago situated about 30 km offshore from southern Brittany. The species is also found in Spain and Portugal but by the beginning of the 20th century the Breton form was believed to be almost extinct. Since 1984, management of the vegetation at a nature reserve at Saint-Nicolas-des-Glénan has helped to increase the population of wild daffodils from 6,500 in 1985 to more than 140,000 in 2003.

Another Breton plant benefiting from the care of the Conservatoire at Brest is the sea lavender, *Limonium humile*. In France this species has never been very abundant and is now known only from the Brest area. Its survival is threatened

The Conservatoire Botanique at Brest and Nantes Botanic Garden are working on a project to conserve and reintroduce *Angelica heterocarpa*, an estuarine species native to France.

The Crozon peninsula in Brittany is one of France's spectacular wild areas. The Conservatoire Botanique at Brest studies and helps to protect threatened plants here.

# A NETWORK FOR FRENCH-SPEAKING BOTANIC GARDENS

A national network of French botanic gardens was created in 1979 to encourage collaboration and exchange of information. In 1996 the network was extended to all French-speaking countries and it is now known as Jardins Botaniques de France et des Pays Francophones (JBF). The association has over 180 members at both individual and institutional level and includes around 60 botanic gardens within France.

The network has developed a charter that defines the principles and practices that a modern botanic garden should follow in order to participate fully in international strategies for plant conservation. To adhere to these ambitious guidelines a garden must comply with a series of rules and duties which include actions in science, research, conservation and education as well as respect for national and international rules concerning access and diffusion of plant material. In 2009, 22 botanic gardens had agreed to abide by this charter. JBF works with the Conservatoires Botaniques so that the two networks combine French botanical expertise into a unique force for plant conservation and public awareness both at home and, increasingly, overseas.

JBF is currently working to improve *ex situ* collections so that botanic gardens in the network can participate more effectively in conservation. The network promotes propagation and exchange of plants between collections to prevent erosion of genetic material, good scientific documentation of collections, development of horticultural expertise and access to collections for research and education programmes.

Cottongrass, a distinctive plant of the Breton heaths.

by the expansion of the invasive North American saltmarsh grass *Spartina alterniflora*. Botanists are trying to find ways to halt the spread of the grass and have ensured that the native *Limonium* is safely stored in the seedbank at Brest and in the living plant collections so that it might be re-planted in the wild.

## OCEANIC ISLANDS

The international work of the garden at Brest concentrates on conserving the threatened floras of oceanic islands. The floras of these islands usually harbour endemic species that are naturally rare and particularly vulnerable to the impacts of habitat disturbance, competition from invasive plants, grazing by introduced animals and, increasingly, the effects of climate change. At the beginning of the 21st century, led by Lesouef, Brest decided to take action to prevent the imminent extinction of the world's most endangered plants, concentrating on those that are considered to be Critically Endangered by IUCN. Stéphane Buord, who is charge of the garden's international work, believes that 'Brest, and other French botanic gardens, should devote their efforts to preserving threatened species, not only locally, but also on an international scale, through cooperation and partnership linkages.'

Rescue operations are underway in collaboration with local partners on a number of islands. Currently, over 1,000 extremely rare species are cultivated in the Brest greenhouses as an insurance policy against loss in the wild and the seed bank has 1,100 species. Sadly, 40 of these plants are already believed to be extinct in their natural habitats. Models developed at Brest to use stored seeds or cultivated plants for restoration of wild habitats have exciting potential. *Ruizia cordata*, a small tree endemic to Réunion was virtually extinct in the wild when it was successfully propagated at the Brest garden and reintroduced to its tropical habitat.

Lesouef made an unsuccessful search for *Ruizia cordata* during a visit to Réunion in 1975, but local botanists found about 14 scattered individuals soon after. In 1977, cuttings were taken from a wild plant and established in cultivation. Most of the wild trees subsequently died after their bark was removed or they were damaged in other ways, probably for use in ritual magic. Cuttings were grown in heated greenhouses in Brest and in 1986 and 1987 two

*Ruizia cordata* has been rescued from the brink of extinction by French botanists but vigilance is still required to protect plants in the wild. Hand pollination of the flowers of this species helped to ensure its survival.

trees flowered. Fortunately, one plant was female and the other male so it was possible to pollinate the flowers by hand and produce viable seed. The following year, 280 plants were returned to Réunion. One hundred of these were planted in inaccessible localities to protect them from collection and the rest were planted in local gardens.

*Normania triphylla*, a conservation success story. Long thought to be extinct it is now safe in cultivation and has been returned to its natural habitat in Madeira.

The island of Madeira has been a particular focus for the collaborative conservation efforts of Brest. The flora of this small rugged volcanic island includes around 130 native plant species that are not found in the wild anywhere else in the world. With the Madeira Natural Park authorities, staff at Madeira Botanical Gardens monitor species and habitats in need of preservation. The botanic garden also works with the Conservatoire in Brest to increase the chances of survival of Madeira's most threatened plant species.

*Normania triphylla*, a plant of the Solanaceae or potato family, was for around 200 years presumed to be extinct. This species, endemic to Madeira, was rediscovered by a local botanist in 1991 and the few seeds collected were shared between the Madeira Botanical Garden and the conservertoire in Brest. Removed from the threat of a plant disease that was rife on the island, the *Normania* plants thrived at Brest and after a few years produced several million seeds. In July 1998, the team at the conservatoire realized the potential of a reintroduction operation and partnered with the Madeira Botanical Gardens and Natural Parks service of Madeira. Seeds were also entrusted to the seedbank at Kew.

*Cheirolophus massonianus*, an attractive Madeiran plant in the daisy family, is considered to be Critically Endangered. It is threatened in the wild by flower picking and grazing, and some time ago botanists had given up hope for the survival of the species in its native habitats. In 1995, botanists from the Madeira Botanical Gardens rediscovered a small wild population on the island of Porto Santo and in 1998 a botanist from Brest found 20 wild specimens on the same island. Seeds collected from these individuals were grown on into new plants at Brest as well as in Madeira. Scientists hope that building up the populations in cultivation should allow the species to be reintroduced to its natural habitats. However, the threats to wild plants from flower picking and overgrazing may remain and so the *ex situ* collections provide a safety net for the species.

Another endangered species that the Conservatoire at Brest has helped to save is the Madeira geranium, (*Geranium maderense*). In the wild, this stunning species grows in the forests of laurel which originally clothed Madeira. Loss of forest habitat and overcollection have taken their toll on the geranium and it is not certain whether any wild populations remain. As the species hybridizes easily with other geraniums, the Conservatoire

cultivates *Geranium maderense* in isolation to be sure of obtaining genetically pure seeds. Fortunately, the species is easy to propagate from seed so it may be possible to reintroduce the Madeira geranium back into the wild.

Left: *Cheirolophus massonianus* is being conserved in *ex situ* collections with a view to reintroduction to its native habitats in the future.

Right: *Geranium maderense* grows well in cultivation but may be extinct in the wild. Options for reintroduction are being considered.

## RÉUNION

The tropical island of Réunion is the largest of the volcanic Mascarene Islands. Its neighbours in the group are Mauritius and Rodrigues. Mauritius passed from French to British control in 1810 and, with its dependency Rodrigues, became an independent country in 1968. Réunion meanwhile remains politically a part of France. Much of the island is mountainous and so more of the native forest cover remains than on Mauritius, where extensive sugar plantations were established in the 18th century. Nevertheless, very little lowland coastal forest remains and the montane forests have been invaded by exotic species. Logging and collection of fuelwood have also degraded the forests. Sadly, three-quarters of Réunion's

## EUPHORBIA STYGIANA

An endangered species of spurge, *Euphorbia stygiana,* originally from the Azores archipelago, was cultivated for the first time in the world at Brest. The ornamental qualities of this plant soon attracted attention: impressive evergreen leaves which have a stroke of white towards the crown and sometimes develop a beautiful, bright red tint towards the end of winter. The small flowers have a delightful smell. *Euphorbia stygiana* is now grown in a number of botanic gardens and is available in the nursery trade.

**three-quarters of Réunion's native orchid species have been lost as a result of agricultural development**

160 or so native orchid species have been lost as a result of agricultural development.

The Conservatoire Botanique on Réunion (known as the Conservatoire Botanique de Mascarin) was established in 1987. A 12.5 hectare garden located close to the town of Saint-Leu, the Conservatoire originally concentrated on *ex situ* conservation and successfully rescued about 60 per cent of the threatened plant species of the island, establishing them in cultivation along with species from the neighbouring islands. Now, the conservatoire is more involved in the management and monitoring of species and populations in their natural habitats. Permanent plots have been established in mid-elevation wet forests for long-term study of plant diversity and vegetation dynamics. Field surveys are carried out to identify, map and monitor locations for rare endemics such as *Ruizia cordata*, which remains highly endangered despite the reintroduction efforts. Areas of high conservation value are assessed for their protection or sustainable management by local authorities.

An important part of the work of the Conservatoire is studying and controlling the invasive plants which are one of the major threats to the native flora. One of the Conservatoire's success stories is the conservation of the endemic *Lomatophyllum macrum*, known locally as mazambron marron. This species, which looks rather like an aloe, is considered to be Vulnerable in the wild and is legally protected in Réunion. It grows naturally in semi-dry forest but because of habitat destruction and competition from invasive species such as the pretty but pernicious shrub, *Lantana camara*, and small trees, *Eriobothrya japonica*, *Rhus longipes*, and *Schinus terebinthifolius*, only about 30 small populations remain. Studies on the reproductive biology of mazambron marron have shown that reinforcing the scattered populations with propagated individuals would encourage higher levels of seed production and increase the chances of successful breeding.

In 2001, almost 300 mazambron marron plants were introduced to two sites which had been cleared of invasive plants, one in a remnant area of native dry forest and the other in a disturbed area of more humid forest where a wild population had been discovered that same year. Seedlings were propagated from seed carefully selected from wild populations to reflect genetic diversity, and plants from the nursery were added to the wild populations to increase their chance of breeding success.

As Maïté Delmas, coordinator of JBF, points out, 'France's efficient network of Conservatories Botaniques is providing a safety net for the threatened species of the country and its overseas territories. The large network of botanic gardens and arboreta, with their extensive collections and skilled staff, is participating actively in ex situ conservation as well as developing educational programmes to increase public awareness of the importance of conserving the world's plant diversity.'

Maïté Delmas, coordinator of Jardins Botaniques de France et des Pays Francophones. She is a passionate supporter of botanic gardens in France and around the world.

Germany has a rich heritage of botanic gardens, with a total of around 100 located throughout the country. The Berlin-Dahlem Botanic Garden is one of the largest and most impressive. Its design and development have had a major influence on botanic gardens around the world. A small botanic garden was first established in Berlin in the 17th century and moved to the suburb of Schöneberg in 1679. The current garden at Dahlem was established in 1910 under the directorship of Adolf Engler, an influential botanist whose views on plant taxonomy and geography have made a lasting impression on plant sciences. The main reason for the move to the present site was to accommodate the huge diversity of plant material collected to realize Engler's vision of a 'world in a garden'.

**Previous page:** The Berlin-Dahlem Botanic Garden is rich with interest, with historic glasshouses, meadows with native plants and geographical collections of plants from around the world.

Many of the plants in the Berlin Botanic Garden are of international conservation interest, with unique ex situ collections of plants from the Mediterranean, Caucasus, Yemen and Cuba. The current director, Dr Thomas Borsch, considers that conservation is now one of the main guiding functions of the garden: 'We have inherited an extraordinary plant collection because of Engler's passion for plants and a responsibility to care for the species that have become increasingly threatened in the wild since his lifetime. All our plants are carefully documented and the methods of propagation recorded so that we can increase the number of individuals of very rare plant species and share our experiences with others. Furthermore, collections of living plants are an important source for the scientific research that increasingly supports conservation activities.'

The garden retains its traditional layout emphasizing plant geography and ecosystem types. An extensive section covering nearly a third of the garden's total area of 42 hectares is taken up by representations of different habitats. Visitors can walk around Northern Hemisphere habitats from the Alps to the Himalayas and on through to Japan and North America. Twelve rock gardens

feature plants from mountains, forests, dunes and steppes, all reflecting Engler's original designs. Also included in the site is a 17 hectare arboretum with around 1,800 different species of tree and shrub from around the world. The whole site is protected as a part of Berlin's cultural and historical heritage.

The species-rich meadows within the garden are more natural, representing the local flora rather than plant-rich habitats from elsewhere in the world. The grasslands, managed in a traditional way and essentially unchanged for around one hundred years, have an abundance of flowers such as meadow clary (*Salvia pratensis*), fairy flax (*Linum catharticum*) and cuckoo flower (*Cardamine pratensis*).

The garden provides a safe haven away from the pressures of urban development and modern agriculture and around 400 plant species grow in the meadows, including 70 that are included on national or regional Red Lists. Examples include the orchids *Listera ovata*, *Epipactis helleborine* and *Platanthera bifolia* together with the maiden pink (*Dianthus deltoides*), meadow saxifrage (*Saxifraga granulata*) and spiked speedwell (*Veronica spicata*).

The bee-pollinated meadow clary is now rare in parts of Europe, such as the UK, but flourishes in the Berlin-Dahlem Botanic Garden.

## LOOKING AFTER LOCAL PLANTS

The city of Berlin and the surrounding area has a high diversity of natural habitats and species. Eighteen per cent of the area is forested, and there are also dry sandy areas and a range of wetland types including bogs, rivers and lakes. In total there are 38 protected Nature Conservation Areas and 52 Landscape Protection Areas. The Berlin-Dahlem Botanic Garden has worked with local conservation authorities for over 20 years to help boost the chances of survival for rare and threatened plant species.

Over 40 plants that are threatened locally or nationally are being carefully cultivated in the garden as part of a programme of integrated *in situ* and *ex situ* conservation work that involves close cooperation with local nature conservation authorities. Species including the black pasqueflower (*Pulsatilla pratensis* subsp. *nigricans*), yellowgreen catchfly (*Silene chlorantha*), Spanish catchfly (*Silene otites*), alpine rush (*Juncus alpinus*) and spike speedwell (*Veronica spicata*) are being conserved in this way. In recent research projects the genetic structure of plants in *ex situ* cultivation has been compared to plants of isolated natural plant populations within the local Berlin area. This information is used to clarify requirements for *ex situ* cultivation in order to maintain genetic variability of the very rare species. The garden's seed bank stores seeds from the locally threatened plants to increase the chances of successful long-term survival of the species. The seed bank currently stores around 1,900 plant taxa. The focus of seed storage is broadening to include rare plants from other areas of Germany and Mediterranean plants, especially from Greece.

Recently, the success of local plant reintroduction projects carried out between 1989 and 1993 has been evaluated with interesting results. In 2008, all fifteen sites in the Berlin area were checked to see whether the reintroduced plants had survived and whether there was an increase or decrease in population size. Unfortunately, few of the plants had survived, mainly because of changes in ground water level, competition with more robust species (especially grasses) and damage from herbivores. Only one species, *Silene chlorantha*, had thrived. The population of this catchfly increased from the 280 plantlets reintroduced between 1990 and 1992 to around 9,500 individuals in 2008. Dr Albert-Dieter Stevens, Head Curator of Living Collections and Herbarium at Berlin reflects that, 'We have learned that long-term monitoring and management of the habitats chosen for reintroduction of rare plants is a necessary prerequisite for the success of such actions. This should be taken into consideration right from

*Armeria alpina* occurs quite widely in Europe but its subspecies *purpurea*, which grew near Lake Constance, is now extinct in the wild.

the start of reintroduction projects. Our work also underlines the importance of *ex situ* living collections and seeds banks in botanic gardens, when the survival of the plants *in situ* can not be guaranteed.'

## CONSERVING NATIVE SPECIES

Botanic gardens throughout Germany are working together to secure the future for threatened native species. This work has been given renewed impetus by the Global Strategy for Plant Conservation. At the time of writing 35 botanic gardens are running *ex situ* conservation projects for native plants. A special plan is in place for the 65 Critically Endangered or very rare plants that are considered to be the highest priority. Thirty-eight of these special plants are being grown in one or more botanic gardens and volunteer facilities and 23 more species will soon be cultivated in the same way. The goal is to have each species in three collections to minimize the risk of accidental loss at any one site.

One rarity being rescued by the national collaboration is *Armeria alpina* subsp. *purpurea*, endemic to the area around Lake Constance and now believed to be extinct in the wild. Offspring of a lakeside population has been cultivated in the Bern Botanic Garden, Switzerland, and is now propagated in the Botanic

Garden of the University of Konstanz for potential reintroduction. Populations of the extinct flax weeds *Cuscuta epilinum*, *Silene linicola* and *Lolium remotum* are cultivated in a specially designed flax field in the Bonn University Botanic Gardens.

In the federal state of Brandenburg, a project has been carried out to help a population of the pink *Dianthus gratianopolitanus* var. *sabulosus*, one of only three surviving wild populations worldwide. A variety found in open pine forests, the natural habitat has been carefully managed to enhance its chances of survival. This wild population has been strengthened with plants propagated *ex situ*, both with seeds sown directly at the forest site and with young plants that were grown in the Heidegarten Langengrassau, a small regional botanic garden.

A young plant of the Critically Endangered cycacd, *Microcycas calocoma* growing in the Pinar del Rio Botanic Garden, Cuba.

## THE FLORA DE CUBA

Berlin's international work reflects collaborations on the study of plant taxonomy and biogeography since the days of Engler and the more recent pressing worldwide need for plant conservation. Behind the scenes, botanical staff at the garden carry out a full research programme, including collaborative projects with countries as diverse as Cuba, El Salvador, Greece and Yemen. The extensive glasshouses within the main tropical greenhouse, one of the largest in the world, allow a wide range of species to be grown for research and display purposes.

Built between 1905 and 1907, and recently renovated, the Great Tropical Greenhouse now has around 1,300 species, most of which are of documented wild origin. The greenhouse includes a display of plants from Cuba, including the rare endemic palm *Coccothrinax borhidiana*.

The association between Berlin and Cuban botany goes back many years. Cooperation is based on a treaty between the governments of Cuba and the German Democratic Republic developed in 1968. Based on this treaty the Cuban–East German joint project on the Flora de Cuba was launched in 1975. German partners of this project were the Botanical Institutes in Jena and East Berlin. In 1989, with the collapse of the Berlin wall, the nature of the botanical collaboration changed. At the end of 1993, the herbarium, remnants of the

## PINAR DEL RIO

The Pinar del Rio region of Cuba has a very rich flora with an estimated 500 endemic species growing in a spectacular karsitic limestone landscape, succulent and thorny scrub and coastal lagoons and mangroves. A botanic garden has recently been established in the region with support from a private organization called the 'Freundschaftsgesellschaft Berlin-Kuba' (Berlin–Cuba Friendship Association). It is anticipated that the collaboration between Berlin and Havana Botanic Gardens will ultimately extend to the botanic garden of Pinar del Rio and others in Cuba.

One particularly interesting plant of Pinar del Rio is the endemic cycad, *Microcycas calocoma*. Known locally as the cork palm, this species is Critically Endangered in the wild. Clearance of the dry tropical forest habitat of this species has been a major threat and scientists believe that the beetle pollinator for the species is now extinct. *Microcycas calocoma* is a flagship species now in cultivation at the Pinar del Rio Botanic Garden, the Havana Botanic Garden and other botanic gardens around the world.

Staff at the Pinar del Rio Botanic Garden are helping to monitor populations of other threatened species in the wild. In recent years their work has led to the rediscovery of a range of species believed to be extinct including the endemic bladderwort, *Pinguicula cubensis* and two species of *Xyris*.

living collection (around 200 species), and staff (scientists and gardeners) moved from the botanical institute in East Berlin to Berlin-Dahlem. Collaboration has intensified, with the first volume of the Flora appearing in 1998 and a further 14 volumes published.

Berlin-Dahlem has established a new official cooperation with the Cuban National Botanic Garden of the University of Havana based on a treaty between the two institutions. This will promote scientific and technical cooperation, including training of gardeners and work on plant endemism and conservation genetics.

Cuba has a very rich tropical flora with 7,000 vascular plant species of which 50 per cent are endemic to the island. The main threats to the flora are habitat loss, fires, agricultural and forestry development and mining. Recently 1,414 plants were evaluated using the IUCN Red List categories and criteria, including 1,089 endemic species. Of the recorded endemic species 21 are believed to be extinct, and 1,006 are threatened with extinction.

## THE CAUCASUS

Work in the Caucasus region also continues a long German tradition of studying the rich flora of this global biodiversity hotspot, estimated to include 6,350 species of higher plants of which around 2,500 are endemic. The Colchic forests, confined to the Caucasus, are temperate rainforest of immense botanical interest. These forests and the rich higher-level alpine flora are considered to be very vulnerable to the impacts of climate change because of their restricted altitudinal range.

The Caucasian flora is rich in timber species, wild crop relatives and species with ornamental potential. Endemic species of wheat, barley, apple and pear are found here, some of which, like Demetrius's pear (*Pyrus demetrii*) are globally threatened. The region is also an important place of origin for the grapevine, with nearly 500 local varieties registered in the country of Georgia. Wine production in the country has a 6,000 year history.

The botanic garden in Tblisi, Georgia has worked closely with German gardens for over one hundred years. The first director of the garden, Adolph

The white-flowered *Rhododendron caucasicum* is relatively rare in cultivation. In the wild it flourishes on northern rocky slopes at an altitude of 1600-3000 metres.

# BERLIN'S HORSE CHESTNUT: SYMBOL OF COLLABORATION

In the Mediterranean section of Berlin-Dahlem Botanic Garden grows a small horse chestnut tree. It is one of the few plants of this commonly cultivated species that is of known wild origin. The history of the cultivation of horse chestnut has been traced using scattered historical literature, old paintings and herbarium specimens. First cultivated in central Europe in the sixteenth century, the tree was soon introduced to the Netherlands, France and the UK. Knowledge of the native origin of the species was lost until the English botanist John Hawkins rediscovered natural populations in the Pindos mountains two hundred years later. Eighty years after Hawkins's discovery Theodor von Heidrich, director of the Athens botanic garden, visited the mountains and found horse chestnut trees growing in shady ravines. Locals apparently used the fruits to cure horses with coughs. The rare and scattered populations of the horse chestnut tree in the Greek mountains are now well-known and the tree in Berlin has come to symbolize of the ongoing collaboration between the garden and Greek botanists in the study and conservation of the Greek flora.

Roloff, was a German born in Tblisi. He was followed by a succession of horticulturists invited from Germany to work in the garden. In 1899, at the start of his tenure, Roloff had the foresight to develop a collection of the unique Caucasian flora within the grounds. Now the garden has an impressive collection of locally endangered plants that are being propagated, cultivated in display gardens and stored in the seed bank. Unfortunately, Tblisi Botanic Garden, in common with botanic gardens throughout the troubled Caucasus region, has suffered periods of neglect, scientific isolation and financial hardship. Berlin-Dahlem Botanic Garden, along with the Royal Botanic Gardens Kew, Missouri Botanical Garden and BGCI, is supporting the redevelopment of Tblisi Botanic Garden. Berlin's initiative has a scientific focus as part of a wider Caucasian plant conservation strategy.

Like all botanic gardens, Berlin-Dahlem has to manage a range of major and at times conflicting goals, including scientific research, maintenance of historically and culturally important garden landscapes, public education and conservation of species, locally and internationally. By working as part of a network and fostering carefully selected international partnerships, Berlin-Dahlem intends to maintain and expand its role as a botanical ark whilst retaining the cultural and scientific heritage that began with Engler's vision.

Nezahat Gökyiğit Botanic Garden, Istanbul

Turkey is at the crossroads of Europe and Asia, and its flora has elements of both east and west in a rich mix that is both precious and precarious. The country has a huge and varied range of landscapes and vegetation and parts of two global biodiversity hotspots, the Mediterranean and the Caucasus, fall within its borders. The flora comprises over 9,500 taxa, one third of which are endemic, and more are still being discovered. Some estimates suggest that a new species is recorded from Turkey every six days. In an ambitious undertaking, Turkey was the first country to document sites that are priorities for plant conservation and by 2007, 144 'important plant areas' were recognized. Protecting these sites is a major undertaking. Threats to the flora include the all-too-familiar loss of natural habitats from agricultural intensification and urbanization, as well as over-exploitation of individual species.

**Previous page:** The natural habitat of *Centaurea tchihatchewii*, one of Turkey's Critically Endangered plants. This species is now in cultivation and restoration of wild populations might be possible.

The Turkish flora has been the source of plant material for the gardens of northern Europe for many centuries. Many familiar garden bulbs, for example, originate from Turkey's rich treasure house of wild tulips, lilies, snowdrops, anemones, iris and cyclamen. The crown imperial fritillary excited the attention of Emperor Ferdinand I of Austria in the 16th century. At around the same time, tulips were first seen in western Europe when they were sent from Turkey to Vienna in 1554. They were considered exotic delights and the bulbs commanded extraordinarily high prices. Ferdinand employed the renowned botanist Clusius, who became Director of the Botanic Garden at Leiden. Although Clusius did not visit Turkey, he is credited with starting the Dutch bulb industry. Tulip mania reached its peak in 1634–1637 when speculators risked everything on rare bulbs in the hope of vast profits.

Over the centuries, huge quantities of wild bulbs have been harvested for export from Turkey, and only in recent years has the export trade been monitored and controlled. Many bulb species are in particular need of conservation attention because their small populations are scarce. At least one Turkish tulip, the late-flowering *Tulipa sprengeri*, is believed to be extinct in the

*Iris nezahatiae.* This species was first described scientifically after it flowered in 2005. In the wild it is known only from a few localities around Yusufeli where it is Critically Endangered as a result of the construction of a new dam.

wild. This popular garden plant with bright red flowers was first grown in western Europe over 100 years ago. Threatened bulb species include 11 species of colchicum, 4 species of grape hyacinth, 8 species of fritillary and 10 crocuses. Most of these have very limited distributions and are now facing extinction.

Some of the richest wild bulb habitats are now legally protected, such as the Spil Dağı (Manisa Dag) National Park in the Aegean region of western Turkey. This area is particularly important for wild tulip species, and it is thought to be the origin of European tulips in cultivation. In addition to protecting bulbs in their natural habitats, careful cultivation of documented wild bulbs in botanic gardens provides an insurance policy against the loss of threatened species. In common with other plant conservation programmes around the world, such *ex situ* collections can be a valuable source of propagation material to replenish threatened plant species in the wild.

## NEZAHAT GÖKYIGIT BOTANIC GARDEN

Bulbous plants are one speciality of the Nezahat Gökyiğit Botanic Garden in Istanbul. This remarkable 50 hectare botanic garden is situated in a busy

The bulb frames at the garden are made from old railway sleepers. Over 380 bulbous species are grown here, providing a reference collection as well as protecting the plants *ex situ*. The garden provides a botanical haven in the middle of the rapidly expanding city of Istanbul.

motorway intersection on formerly derelict land in a residential area of Istanbul and is now the largest replanted green area in this huge city. Established in 1995 by wealthy businessman Nihat Gökyiğit as a memorial to his late wife, the park became the Nezahat Gökyiğit Botanic Garden in April 2003. Its mission statement is: 'To explore, explain and conserve the world of plants'. Adil Güner, an internationally respected botanist and conservationist, is the director of the garden and the driving force behind its conservation programme. Güner has travelled extensively throughout Turkey and is aware of the major pressures on the country's plant diversity. Ever since his doctorate on *Iris* he has had a special interest in the genus and irises are well represented at the garden.

Bulbous plants, are, however, just one of the plant groups of interest to Güner. He has developed an extensive collection of over 380 Turkish geophytes (bulbs, tubers, corms and rhizomes) from 36 genera, grown in over 30 bulb frames, which hold pots sunk into raised beds made from old railway sleepers. The bulbs have all been collected from the wild during the past few years and carefully documented to form a valuable conservation and research collection which is used not only for display and for public education, but also for propagation and reintroduction into the wild. Spring visitors can see many bulbs in flower, including galanthus, cyclamen, crocus, narcissus, iris and tulip.

Nezahat Gökyiğit Botanic Garden is being developed in a country where there are relatively few arboreta or botanic gardens in relation both to landmass size and the richness of the flora. One of its conservation missions is to protect endemic and rare plants of the Istanbul area, which are under severe threat because of rapid urban development. It also aims to help conserve Turkey's flora nationally. In developing the garden, extensive advice was sought from the Royal Botanic Gardens, Kew, in particular from Margaret Johnson, and from David Rae at the Royal Botanic Garden Edinburgh. Experts from these gardens regularly visit Istanbul and exchange visits between the institutions have benefited everybody. Sometimes these visits are linked to fieldwork, both to collect living material and to assess the conservation needs of plants in the wild.

## NOT JUST BULBS

As well as conserving bulbous plants, Nezahat Gökyiğit Botanic Garden is working on a range of other species. *Centaurea iconiensis*, a yellow-flowered knapweed, is considered to be critically endangered in Turkey. As only 22 individuals were known to survive in the wild in Central Anatolia, botanists decided to take action to increase the population and save the species. Seeds from this attractive plant were successfully cultivated in the garden as the basis for developing a recovery programme. With the help of local villagers, two areas were fenced off in its native habitat and most of the seedlings grown in the garden were transplanted there. During transplantation activities and celebrations, the villagers told botanists from the garden about a new, previously unknown population. Now the villagers are aware of the rarity of the species and are looking after the plants. Botanists are helping to monitor the wild populations to ensure the long-term survival of the species.

A similar programme was started for the wild pear *Pyrus serikensis*, a species that is also Critically Endangered. It grows near to Antalya where villagers eat

*Centaurea iconiensis* is one of Turkey's rarest plants. Fortunately there is an active conservation programme to save this Critically Endangered species from extinction.

the small pears. Trees are used for grafting and it is also a useful shade tree. Tragically, during the hot summer of 2008, forest fires swept through the area near Antalya for several days, destroying many thousands of hectares of forest, including 10% of the area where the pear was known to grow. A total of 1,067 *Pyrus serikensis* trees were counted in a specially protected area near Antalya, which covers about 50 per cent of its total distribution.

About 400 seedlings were already being grown in the garden, but since the disaster many more seeds have been collected with the aim of growing more plants for reintroduction programmes. With the support of the Turkish government, garden staff have transplanted these seedlings into the burned areas in an effort to recover population numbers.

It is hoped that similar steps can be taken to restore another Critically Endangered species, *Centaurea tchihatchewii*. This attractive red-flowered annual is found in central Anatolia, where it is known from only two populations. One of these was under threat due to the rapid urban development of Ankara and is already surrounded by the city development; it is now protected. In the garden the species is well established and produces flowers from December through to May.

*Amsonia orientalis* is the subject of another successful conservation project at the garden. 'Blue Star' is well known in cultivation, but in the wild it is known only from a few areas near Istanbul and around the Sea of Marmara, where it grows in damp shady places.

Other species being shortlisted for conservation projects include *Astragalus beypazaricus,* known from just one weak population near the main road between Beypazari and Nallihan west of Ankara. Local transport authorities have agreed not to widen the road in this area because it would destroy these vulnerable plants. The major challenge with *A. beypazaricus* is that it doesn't set fruit in the wild and propagation efforts have also failed. A second *Astragalus* species, *A. yildirimlii,* is also being conserved. It is known from one locality near Beypazari, and is also under threat, although the population is healthier than that of *A. beypazaricus.*

The wild pear, *Pyrus serikensis*, which grows near Antalya, is considered to be Critically Endangered as a result of habitat destruction. Fire has been the most significant recent threat. Action is now being taken and there are plans to make the species a flagship for local conservation.

## BOTANICAL EDUCATION

Educating the public about the importance of plants and the need for plant conservation is vital if further plant extinctions in Turkey are to be avoided. As well as guide books in Turkish and English, a colourful gardening magazine, *Bagbahçe*, provides a platform for news and information about the garden along with other related articles. An illustrated children's booklet *Bitkileri Taniyalim* (*Let's learn about plants*), together with a teachers' pack, has also been published.

Nezahat Gökyiğit Botanic Garden is pioneering education at all levels – from local school children to government officials. A children's gardening project which began in 2008 offers small groups of local children the opportunity to grow and maintain their own vegetables and flowers. Along the way the children learn about plant and water conservation, healthy eating, recycling and compost making. Garden staff hope to expand this approach in the future with gardens at schools. An area of the site has been landscaped as an exciting play area where children of all ages can learn about the importance of plants, and where their food and everyday items come from. For children growing up in a city as huge as Istanbul, where there is less and less contact with the village life that their parents or grandparents knew, this is particularly important.

Children are encouraged to learn about nature and the rich flora of Turkey through play and special publications.

As a part of several education programmes, a school has 'adopted' the endemic *Fritillaria carica* subsp. *serpenticola*, which is Critically Endangered in Antalya province. This dainty yellow spring-flowering fritillary is known in the wild from only one small population where it grows on serpentine rocks. The children followed its life cycle throughout the year, drew pictures of the plants, and sold badges and T-shirts at their school fair, helping to publicize the plant's vulnerability.

With the aim of showcasing plants that tolerate arid conditions, an ambitious garden area for dry and halophytic (salt-tolerant) plants has been established. There is also a medicinal, aromatic and economic plants garden and a collection of dye plants. Throughout the garden, informative labels explain to the thousands of visitors the importance of plants and their many roles and uses in everyday life.

None of these projects, including the garden itself, would have been possible without the sponsorship and unfailing enthusiasm of the founder of the Garden, Nihat Gökyiğit. This remarkable man has not only been a constant inspiration for the rapid development of this unique project, but he has also spread the word about this unusual botanic garden and the importance of

New plant species continue to be discovered in Turkey including this yellow-flowered *Iris*.

Turkey's rich biodiversity far and wide. He was cofounder of the TEMA Foundation (the Turkish Foundation for Combating Soil Erosion, and the Protection of Forests and Natural Habitats) in 1992 and more recently he has produced booklets about the biodiversity of Turkey in both Turkish and English. He is involved in many social and environmental campaigns and has received many awards.

# EDINBURGH AND THE FLORA OF TURKEY

The Royal Botanic Garden Edinburgh has had a long-standing connection with Turkish botany through the *Flora of Turkey* project. From 1965–1985, under the leadership of Peter Davis, Edinburgh took the lead in cataloguing the flora, publishing nine volumes. A supplement was published in 1988 and the leadership of the project was subsequently handed over to Turkish botanists, with a second supplement published in 2000.

Adil Güner remembers his first three-month visit to Edinburgh in 1982 with affection (not least because it was the first time he had left Turkey, and the journey from Istanbul to Edinburgh by train took several days). At that time his English was very limited and he found it very difficult to understand the language and especially

Adil Güner shows colleagues from Edinburgh and Kew the bulb collection at NGGB.

Scottish accents. However, the experience of seeing such a famous botanic garden, meeting Peter Davis and Ian Hedge and working in the herbarium was wonderful.

It was to be the first of many visits and marked the beginning of a long collaboration. Handing in the final manuscript for Volume 11 of the *Flora of Turkey* to the Edinburgh University Press was a particularly proud moment for Güner: all previous volumes had been written by botanists based in Edinburgh, so Volume 11 had a special significance because it was the first to have been written in Turkey.

Three years later, Güner was back in Edinburgh and over a dinner with Nihat Gökyiğit, Margaret Johnson and the Regius Keeper, Professor Stephen Blackmore, the seeds were sown for resurrecting the collaborative links between Turkey and Scotland. This marked a historic turning point, since in the past material had been taken from Turkey. Now the tables were turned; this time the British garden would be actively helping Turkey.

2005 saw these links further strengthened with the start of a very successful three-year project overseen by Edinburgh's Director of Horticulture, David Rae. Funded by the UK Government's Darwin Initiative, a scheme that assists countries rich in biodiversity but poor in financial resources, the project involved an exchange of staff between the two gardens with the important aim of strengthening horticulture and education for conservation work. In June 2008 the two gardens agreed to ensure that exchange visits between the two institutions would continue in the years to come.

# EGE UNIVERSITY BOTANIC GARDEN

The Ege University Botanic Garden, near Izmir in western Turkey, was founded in 1964. A central aim of the garden's research centre is to protect the endemic endangered plants of Turkey through scientific studies on the flora and to share the data obtained with scientists and the general public.

The facilities include greenhouses, a display garden, systematic beds, an arboretum, an orchard, pools and a herbarium comprising 4,1000 specimens. Over 2,000 taxa from different habitats are on display and students of all ages visit the garden to learn about the world's plant diversity.

*Fontanesia philliraeoides*, a member of the olive family confined to the eastern Mediterranean, is one of the species protected in Turkey's botanic gardens.

Since 2000, studies have focused on conservation biology. Fifty endemic taxa are being protected in the garden *ex-situ* and since 2007 the Bengisu Conservation Garden has been in development 100 km from the university. Along with the garden at Ege, other established and developing botanic gardens in Turkey are now working towards creating a network to help meet the challenges of protecting the country's threatened flora.

## THE FUTURE

Nezahat Gökyiğit Botanic Garden is keen to expand the conservation role of the 13 botanic gardens and arboreta in Turkey by sharing experiences and encouraging networking. The first Turkish National Botanic Gardens Symposium was held at the Aegean University Campus in Izmir in May 2005 and marked the 40th anniversary of the founding of Ege University Botanic Garden. The meeting focused on the current situation of both established and newly developing botanic gardens, their foundation, ideals, functions and working principles. Attendees discussed the challenges of financial and funding difficulties, technical problems and the lack of qualified staff and concluded that a network should be set up to improve communication and management of national collections of Turkey's rich biological hertiage. Discussions for starting a national, government-funded botanic garden in Ankara are now under way, together with plans for two more, one at the Forestry Ministry and another at a university in Ankara.

With just 13 botanic gardens and arboreta but close to 2,000 endangered species, Turkey faces an enormous challenge and the development of botanic gardens is vital. 'We need to act on the information that has been collected by botanists on the rare plants of Turkey and make sure these all survive as part of our natural and cultural heritage,' says Güner. 'Ultimately the rural communities should protect their local plants, but we need to use the technical skills of botanic gardens to help restore the full floral diversity.'

The Fairchild Tropical Botanic Garden is a beautifully laid out, lush garden in Miami, Florida. The garden was established in 1938 by the plant explorer David Fairchild, who travelled the world collecting plants that could be grown in the US. Initially a research and display garden, the biodiversity conservation role of Fairchild has become increasingly important. Under the leadership of former director Michael Maunder, the conservation role became global in outlook. The garden continues to support plant conservation action in Central America, the Caribbean, East Africa and Madagascar, as well as in the local Florida ecosystems. 'The same techniques can be used to save plants from extinction in different parts of the world,' says Maunder. We just need to adapt the necessary activities to local conditions – and take action now.'

**Previous page:** Native pineland vegetation in Florida. This very rare type of Caribbean vegetation is being cleared at an alarming rate.

Within Florida, Fairchild is playing a key role in the conservation and restoration of the pine rocklands, an endangered ecosystem found only in South Florida, the Bahamas and Cuba. The characteristic tree of Florida's pine rocklands is the South Florida slash pine, *Pinus elliottii* var. *densa*. Growing under the canopy of pines is a fascinating mix of tropical and temperate plants, indicating Florida's position as a crossroads for plant diversity. Tropical palms such as the saw palmetto (*Serenoa repens*) grow among a mixture of grasses, sedges and shrubs. Altogether 374 different plant species grow in this habitat type, of which 31 are endemic to Florida.

The Floridian pine rocklands, which occur on relatively high ground on limestone rock, have been reduced to tiny fragments representing less than two per cent of their former area. The relentless pace of building development has eaten away at this unique habitat type. Today, five of the plant species associated with the pine rocklands are on the US government's endangered species list and five more qualify for this category of threat but await formal listing.

One of these is the crenulate leadplant (*Amorpha herbacea* var. *crenulata*), a dainty-flowered plant that is only found in Miami-Dade County. In the past 10

years, habitat destruction has wiped out over half of the wild population and reduced its range to a handful of reserves. Fairchild is helping to rescue the species, working in partnership with the local land conservation and transport authorities. Over 400 individual plants and several hundred seeds have been planted in two separate reserves in an experiment to see whether seedlings, cuttings grown in the nursery, or whole plants rescued from the wild will thrive in a new location. At the same time, the precise ecological preferences of the leadplant are being studied to maximize its chance of long-term survival.

Another endangered pine rockland plant benefiting from Fairchild's care is the coral hoary pea (*Tephrosia angustissima* var. *corallicola*). Without intervention the likelihood of this species surviving is low, as it is known only from a single population in a cultivated field. Fairchild has propagated the coral hoary pea within its nursery and is now testing the effect of three natural microhabitats on growth, survival, and population viability.

Dr Joyce Maschinski, Fairchild's conservation ecologist, feels a great sense of responsibility for the survival of this plant. 'It would be easy to overlook the coral hoary pea, but we cannot afford to lose this plant. Our local plant diversity is ecologically important and helps define Florida's natural identity. Over the five years of our study, patches of the coral hoary pea have established in moist sections. We are beginning to see second-generation seedlings that give us hope that the population is sustainable.'

The goatsfoot passionflower (*Passiflora sexflora*) is one of 53 passionflower species native to the US. This species is also distributed in various countries of the Caribbean and Central America. Fairchild has worked with various local groups and a private landowner to help restore this species to Florida, where it occurs only rarely in the pine rocklands. The reintroduction programme has included detailed fieldwork to understand the distribution and ecology of this species, the collection of wild plants, nursery propagation, and reintroduction

of over 100 plants to locations within the passionflower's historic range. The results have been impressive. Reintroductions have tripled the known population size for the passionflower in Florida, with a 90 per cent survival rate for transplants and over one-third of these plants reproducing.

**Left:** The goatsfoot passionflower is one the species rescued by scientists working at Fairchild Tropical Botanic Garden. Following propagation in the garden, plants were carefully returned to the wild and this species is now reproducing well in its native habitat.

**Right:** The coral hoary pea in cultivation at Fairchild. Without intervention the likelihood of this species surviving is low.

## HABITAT RESORATION

Rescuing and replanting the endangered plants of the pine rocklands is like fitting back together pieces of a jigsaw puzzle. Fairchild is also looking at the bigger picture in habitat restoration. The goal is to create corridors between the fragmented pine rocklands by replanting a wide mix of the native plant species to connect the fragments. Ultimately, this will benefit overall biodiversity as well as the endangered plant species. Rare and threatened animals that are associated with the pine rocklands in South Florida include the endangered Key deer, (*Odocoileus virginianus clavium*) a subspecies of the white-tailed deer found only in the Florida Keys; Kirtland's warbler (*Dendroica kirtlandii*), bald eagle (*Haliaeetus leucocephalus*), and five endemic reptiles. All these and the rich invertebrate fauna ultimately depend on the local diversity of plants for their survival.

# THE FLORIDA ATALA BUTTERFLY

The Florida atala butterfly, (*Eumaus atala*) is found in
Florida, the Bahamas and Cuba, where its natural habitats
are pine rocklands and tropical hardwood hammocks (areas
of dense forest with a closed canopy of hardwood tree
species). In Florida the butterfly is considered to be
endangered mainly as a result of habitat loss and over-
collection of the cycad *Zamia pumila*, which is the food
plant of the caterpillar. The cycad is also endangered in the
wild. It has in the past been used by Native Americans as a
food plant, but required careful preparation to destroy
toxins. These poisonous compounds are absorbed by the
Florida atala caterpillars and form part of their defence
from predators.

    *Zamia pumila* is now commonly grown in Florida and the
caterpillars of the Florida atala are sometimes considered
to be a horticultural pest, despite their endangered status.
This cycad is grown in Fairchild's impressive cycad
collection – one of the largest collections of cycads on
display in North America. Consequently, the Fairchild
Tropical Botanic Garden is a haven for the butterfly.

The Florida atala butterfly. Endangered in the wild, the butterfly is nevertheless
sometimes regarded as a horticultural pest.

The pine rockland ecosystem is fire-adapted, which means that the species
and species assemblages that make up this forest type are adapted to cope with
natural burning. If the fire regime changes, then so does the species make-up of
the pine rocklands. The suppression of natural fires has resulted in pine
rockland vegetation being replaced by hammock-like plant communities.
Hammock ecosystems are another important habitat type in Florida –
characterized by dense stands of hardwoods such as black cherry, flowering

**Left:** The Key deer is found only in the Florida Keys and is closely associated with pine rockland habitat. It feeds on a variety of plants including palms and mangroves.

**Above right:** Woody hammock vegetation. Hammock ecosystems are an important habitat type in Florida, but they are encroaching on nearby pine rocklands because natural fires are suppressed.

dogwood, laurel oak and live oak. However, hammock plants that invade adjacent upland pine rocklands are normally killed by fire, and the rockland habitat is maintained. Without natural burning, the hammock vegetation spreads, grows more dense and replaces the species of the upland pine rocklands. Maintaining the balance of indigenous species is an important component – and major challenge – of habitat restoration and management.

## PLANT REINTRODUCTIONS

Hammock shrub verbena (*Lantana canescens*) is a rare tropical shrub native to the transitional vegetation between pine rockland and hardwood hammock forests in Florida – a habitat that has almost disappeared. In 2005, Fairchild scientists worked with dozens of volunteers to conduct an experimental outplanting of the species with the hope of preventing its extinction in Florida by understanding its habitat requirements. Initial findings showed that light is very important for seedling production – plants in sites without shading produced plenty of new seedlings within 18 months of planting. This suggests that under natural conditions, the hammock shrub verbena is being edged out by the spread of denser vegetation.

Scientific expertise, appropriate horticultural facilities, and an impressive base of volunteers committed to looking after local plantlife have all been essential components of plant reintroduction programmes in Florida. Over the past 18 years, 32 reintroductions of 11 species have been undertaken, in collaboration with land managers at partner agencies. The conservation collections maintained within the garden are carefully designed and tended to conserve and augment natural populations of endangered and threatened species. Often an experimental approach is needed as each species reintroduction into the natural world is a complex process.

The first species reintroduction undertaken by Fairchild was of the globally vulnerable Sargent's cherry palm (*Pseudophoenix sargentii*). Palms, so often associated with Florida, are a particular speciality of staff at the garden, who are cooperating in palm conservation activities worldwide. Sargent's cherry palm is an attractive Caribbean palm with bright red fruit, which can be seen growing in the Fairchild Palmetum. It has suffered in the wild from over-collection, from general habitat destruction and from the impact of hurricanes. At the time of its discovery in Florida in 1886, this species of cherry palm was known from three islands in the Florida Keys. By 1925 it was thought to be extinct at those locations. Then, in 1950, 28 adult palms were rediscovered on Elliott Key, the northernmost of the true Florida Keys, lying about 12 miles off the coast of mainland Florida and part of Biscayne National Park. In 1991 a thorough survey by Fairchild scientists found that 47 plants remained on Elliott Key and no other wild plants remained in Florida. It was time to take action. Working with the National Park Service and the Florida Department of Environmental Protection, Fairchild's conservation collections were used to supplement the wild population and reintroduce the species back to its three

Students learning how to restore native vegetation under the guidance of Fairchild botanist Sam Wright.

islands. A long-term monitoring programme was put in place to check the progress of the slow-growing and long-lived palm.

Maschinski has been involved in a review of the overall success of this reintroduction programme – one of the best-documented plant species recovery actions in the world. She is pleased to report that Sargent's cherry palm is a conservation success story but says that continued vigilance will be necessary. 'Since 1991 the wild population has increased more than six-fold and our efforts have certainly helped. First, the transplanted individuals contribute to the overall number of plants in the wild – lowering the extinction risk. Second, we have increased the total locations where this plant grows so there is less chance of the whole species being wiped out. As you can imagine, these populations have gotten their share of hurricane impacts!'

**Above:** Young plants of the Key tree cactus (*Pilosocereus robinii*) in the wild and in cultivation. This species has a very restricted distribution in the wild and is subject to intensive conservation efforts as tourism and other forms of development place pressures on the natural vegetation of the Florida Keys.

**Right:** *P. robinii* in its native habitat.

## CONSERVING CACTI

Two other very rare plants of the Florida Keys being conserved by Fairchild are the cacti *Pilosocereus robinii* and *Consolea corallicola*. Commonly known as the Key tree cactus, *Pilosocereus robinii* is a truly magnificent cactus that can grow up to 10 metres (about 30 feet) tall with showy white flowers. This species is found in some botanic garden collections but the only place in the US that the cactus grows in the wild is in the Florida Keys. It is listed by the US government as an endangered species because of its limited distribution and the threats it faces. Some botanists consider this cactus to be the same as the robin tree cactus *Pilosocereus polygonus*, which also grows in the Bahamas, Cuba, Haiti and the Dominican Republic. Whichever classification is followed, the Key tree cactus is an important component of the natural habitat in the Florida Keys and the eight remaining populations need special help. Although some sites are protected, land development continues to pose a risk to

the cactus, and in common with other emblematic cacti like the saguaro of 'Wild West' fame, poachers remain a problem. Additionally, during the past decade, some of the largest populations have experienced a sudden and rapid decline with many plants dying. Now cactus 'skeletons' are a familiar sight and few healthy individuals can be found, especially in the lower Keys. Researchers at Fairchild are working in collaboration with the US Fish and Wildlife Service and with the Desert Botanical Garden in Phoenix, Arizona to try to understand what is causing this decline.

Theories about why the Key tree cactus populations are dying include increased exposure to salt water from storm surges during hurricanes and/or from the rise in sea level due to climate change. Another theory is that tree canopies in hammock vegetation have become denser over time, limiting light availability for the growth of young plants. This theory is supported by photographs of the cacti from around 30 years ago, which show a more open habitat. And another factor to be considered is possible impacts from the Key deer. With help from conservationists, the population of this deer is recovering and has increased from fewer than 50 individuals in the 1940s to an estimated 600–800 in 2000. Several large cacti have been found with wounds near their bases, which may be due to the deer's habit of rubbing their antlers against tree trunks. Unlike hardwood trees, the outer surface of the cactus is fleshy and not protected by tough bark. When the fleshy material is damaged, the cacti cannot transport water and nutrients between the roots and the upper portions of the plant. It seems likely that the decline of the Key tree cactus is due to a combination of factors. When these are better understood, the chances for species recovery will be greatly improved.

Fairchild's research and conservation efforts for the Key tree cactus have involved mapping all the remaining individuals in the Keys, gathering and analysing information about the ecological conditions where they grow, such as soil salinity and canopy cover, collecting material for propagation and reintroduction, collecting seeds for research and long-term storage, and partnering with local agencies to protect populations on private land. Suitable places for reintroductions are being explored. The aim is to boost

Several Key tree cacti have been found with wounds near their bases, possibly caused by Key deer. The decline of the species in the wild is not fully understood and is probably due to a combination of factors. Fairchild scientists are studying the problem so that they can help conserve the wild populations.

the numbers of plants in the wild and find out what habitat-management methods would best enable this impressive cactus to flourish.

The endangered semaphore cactus (*Consolea corallicola*) is known from only two populations of just a few individuals in the Florida Keys and may very well be the most endangered plant in the US. The species is hugely disadvantaged by its low genetic diversity and low rates of reproduction. Genetic studies conducted by Drs Carl Lewis and Javier Francisco Ortega in Fairchild's genetics laboratory confirmed that the two populations are genetically identical, indicating that the individuals are all clones of a single plant. In addition, the species is threatened by an exotic moth, *Cactoblastis cactorum*, that preys on closely related *Opuntia* species. Experimental feeding trials at the US Department of Agriculture, for which Fairchild provided plant material, confirmed that the caterpillars do prey on *Consolea corallicola*. Fortunately, this cactus is in cultivation at botanic gardens, which provides some hope for its future survival. Fairchild holds the Centre for Plant Conservation (CPC) national collection for this species as well as the Key tree cactus, and is collaborating with other regional and national organizations to ensure its survival.

Volunteers play a very important support role in reintroducing threatened species to the wild.

## INVOLVING THE COMMUNITY

Visitors to Fairchild Tropical Botanic Garden cannot fail to be struck by the tranquility and attractive layout of this South Florida garden. The palm, bamboo and cycad collections are hugely impressive with large numbers of endangered species represented, including more endangered palm species than any other botanical garden – and 42 per cent of the world's cycad species, all threatened with extinction in the wild. The rainforest display, initially created in 1959, conveys a sense of natural tropical forest vegetation and is an important educational resource. The Lynn Fort Lummus Endangered Species Garden, incorporated into the pine rockland habitat exhibit in the Jewels of the Caribbean display, provides a permanent exhibit of local endangered plants (Fairchild's collections include 45 per cent of the

introducing students of all ages to plants and expanding their knowledge of botany is an essential component of Fairchild's conservation projects

federally listed and candidate species from South Florida). Against a backdrop of more common native plants, the display includes rare palms, cacti and grasses from Fairchild's conservation collections. This provides an important visual reminder that Florida has more nationally endangered plants and animals than any other US state except for Hawaii and California.

Behind the scenes, beyond the beautiful garden displays, the meticulously executed conservation research and practice is of major importance. But this is not jealously guarded by the scientists. Fairchild is very keen to get all sectors of society involved in plant conservation, and the volunteer programmes are impressive. For younger people, the Fairchild Challenge, set up in 2002, reaches out to schoolchildren and aims to promote environmental awareness in youngsters, through a range of competitive projects. Challenges might include habitat restoration or energy audits; participating schools are awarded points for their work and compete for recognition at the annual Fairchild Challenge Awards. In one example, in

2007, 100 students guided by the South Florida Conservation Team helped with *Amorpha herbacea* var. *crenulata* habitat restoration. In another scheme, Fairchild's South Florida Conservation Team led high school students to two Dade County natural areas, where they learned about the challenges of habitat management, including invasive species, threatened species, fire ecology and habitat fragmentation. The Challenge has been such a success that the idea has spread not only to institutions elsewhere in the US, including zoos and museums, but also abroad to countries such as South Africa and Singapore.

Introducing students of all ages to living collections of plants and expanding their knowledge of botany is an essential component of the garden's conservation projects and has the added benefit of improving ecological awareness among the general public. In a highly urbanized society Fairchild Tropical Botanic Garden provides an 'ecological bridge' to the natural world.

The Chicago Botanic Garden is relatively young in the history of botanic gardens, having opened in 1972 as a centre for plant collections, education and research. Its mission is 'to promote the enjoyment, understanding, and conservation of plants and the natural world'. The garden covers over 155 hectares, more than half of which is native habitat, with woodland, prairie and aquatic ecosystems all represented. In 1997, the garden initiated a research programme on threatened plants, which has grown to become an internationally recognized leader in plant science and conservation research. Staff at the garden are working to secure a future for threatened plants within the American Midwest and beyond.

**Previous page:** Trilliums in bloom in the native oak woodlands of Chicago Botanic Garden.

Dr Kayri Havens has directed the growth of the Division of Plant Science and Conservation at the garden since its inception in 1997. Her motivation stems from a lifelong interest in plants and a desire to prevent the needless loss of plant diversity. 'Having grown up in the Chicago region, the tallgrass prairie has a special place in my heart. I consider myself extremely lucky to have a career that allows me to focus on my passion for plants and the unique Midwestern landscapes.'

The American Midwest historically contained a mosaic of prairie vegetation, interspersed with deciduous woodland habitat. At the time of European settlement, prairie covered about 40 per cent of the US. Grazed by bison and supporting a rich wildlife, prairie habitat ranges from shortgrass prairie in the west, influenced by the rainshadow of the Rocky Mountains, with mixed prairie further east and tallgrass prairie in the moister, more fertile eastern states. Today, less than one per cent of these natural prairies remains, usually as tiny remnants. Agricultural expansion and fire suppression rapidly contributed to the decline of the prairies and urbanization, industrial development, pollution and the introduction of new aggressive weeds have also played a part. Great efforts are

being made to look after the natural prairie fragments and restore degraded areas around Chicago, and the Chicago Botanic Garden is playing an important role in these efforts. Within the grounds of the garden lies the Dixon Prairie, a recreation of six prairie ecosystems that once existed in the area. It is now home to more than 250 prairie plant species and offers trails that provide visitors with opportunities to explore and appreciate the diversity of prairies in the region.

A number of US prairie plants have become popular garden plants around the world, including *Liatris, Aster* and *Penstemon*. The widely cultivated medicinal coneflower, *Echinacea angustifolia*, is a prairie species that has become increasingly fragmented in the wild. Garden researcher Dr Stuart Wagenius is studying the ecological and genetic impacts of this fragmentation on the future survival of wild plants.

*Dalea purpurea*, the purple prairie clover, is one of many attractive native plants that visitors can see on the prairie trails at Chicago Botanic Garden.

## SAVANNAH AND WOODLAND

Along with the prairie grasslands, oak savannah and woodland were found in regions where fire was less frequent. Dominated by species such as *Quercus macrocarpa* and *Q. rubra*, now only fragments of this habitat remain. McDonald

Woods is the Chicago Botanic Garden's naturally occurring oak woodland community. Differences in soil type, moisture and topography have fostered considerable plant diversity within the site. Over 300 plant species support more than 100 species of bird, 20 species of mammal (including the southern flying squirrel, masked shrew and hoary bat) and countless insects. Many of the plants found here are no longer common in the region, and several, including northern cranesbill (*Geranium bicknelli*), forked aster (*Aster furcatus*) and American dog violet (*Viola conspersa*), are either threatened or endangered in the state of Illinois. *Viola conspersa* is on the edge of its range in Illinois and is declining in native woodlands, due to the impact of invasive species such as buckthorn (*Rhamnus cathartica*), a common European shrub

that has become widely naturalized in the US and is considered a noxious weed in some states, and another European introduction, garlic mustard (*Alliaria petiolata*).

Before the Chicago Botanic Garden was established, this patch of woodland was declining as a result of new land uses, invasive species, fragmentation and fire suppression. In an effort to restore the degraded habitat and improve species diversity, the garden now actively manages the woodland through controlled burning, removal of invasive species and reseeding. The dog violet has particularly benefited from the removal of invasives and is now on the increase. Scientists are prioritizing investigations into the impacts of invasive plants and animals on the woodland. Finding out whether the existing restoration methods are addressing all the damage done by invasive species is also a high priority, as is

Controlled burning is necessary to maintain the ecological balance of prairie vegetation.

determining how best to monitor these vital ecosystem processes. Monitoring woodland species and ecological conditions is a continuous process which takes into account the small mammal, bird and insect populations as well as the plant species.

## CITIZEN SCIENTISTS

Chicago Botanic Garden is keen to demonstrate its conservation and restoration work to visitors and involve non-scientists to the fullest extent possible in its work. As part of this commitment, the woods contain marked trails that allow visitors to enjoy the unique remnant woodland, particularly in early spring when the violets and other attractive woodland plants like

# ASTER FURCATUS

*Aster furcatus*, the forked aster, is a globally rare plant that is listed as threatened by the State of Illinois. It is only found in the upper midwest region of the US, where small populations occur in six states, generally growing on alkaline soils in damp woodland areas. Large white flowers bloom from July to October, but they often fail to produce seed. The species usually reproduces asexually through rhizomes that spread underground, eventually emerging up to 40 cm away from the parent plant. Because of this mode of reproduction, populations of the forked aster often consist of clones comprising only one or a few genetic individuals. This small gene pool may reduce the long-term survival prospects of the species by, for example, limiting its potential ability to adapt to the impact of climate change.

In 2009, 19 populations of *Aster furcatus* were known in the Chicago area. Genetic analysis of one population revealed low genetic variability, which may be contributing to the reduction in seed production. Other studies found that an increase in light is associated with an increase in flowering, and future investigations will focus upon the effect of canopy light availability on flowering patterns and the impact of deer browsing. Chicago Botanic Garden is the primary custodian for *ex situ* conservation of this species under the National Collection of Endangered Species scheme. Seeds are stored in the garden's seed bank, and the health of wild populations is monitored as a part of the garden's rare plant monitoring programme, Plants of Concern.

*Trillium grandiflorum* are in full flower. Complementing the garden's natural areas is a more formal native plant garden, which shows visitors how they can contribute to conserving rare habitats by creating a patch of prairie or woodland in their own garden. This garden is located in a prominent position near the visitor and education centres, ensuring that the over 760,000 visitors received by the garden each year are able to connect with the region's plant diversity.

Linked to conservation in the Chicago region, the native plant garden forms a demonstration site for Chicago Wilderness, a regional nature reserve that includes some 90 hectares of protected natural areas, including public and private land. The Chicago Wilderness consortium brings together over 200 public and private organizations who are working to protect, restore, study and manage the region's natural ecosystems and enrich local residents' quality of life. Chicago Botanic Garden was one of the first consortium members and provides a base for the Science and Natural Resources Management Team Coordinator.

Ecosystem management requires an understanding of the component species and how they function. Chicago Botanic Garden is involved in regional floristics – listing and classifying the local flora and providing baseline information for monitoring and conservation. Again involving local people as citizen scientists, the garden runs a monitoring project called 'Plants of Concern', designed to collect population data on the region's rare plants. Since it began in 2000, nearly 400 volunteers have been trained through workshops, courses and field training and more than 500 populations of 180 species have been monitored. The volunteers record GPS locations and collect data on population size and area, phenology (seasonal phenomena, such as flowering time), threats and management activities. Their work helps to track the health of plant populations over time and provides essential conservation data to land managers. This long-term project is part of the Chicago Wilderness Biodiversity Recovery Plan goal to assess, research, protect and recover the region's rare plants.

Echinacea is one of the important species of the prairies that has been used for centuries as a medicinal plant.

Coupled with work on floristics, the Chicago Botanic Garden carries out research on plant ecology and restoration, plant rarity, and plant–soil interactions, both within the garden grounds and in habitats elsewhere in the Great Lakes region. Investigations of constraints on plant conservation, including climate change, fragmentation, invasive species, disturbance and changes in land use are particularly important.

## CONSERVATION CASE STUDIES

One flagship species that is benefiting from the work of Chicago Botanic Garden is the eastern prairie fringed orchid (*Platanthera leucophaea*), a showy, short-lived perennial orchid with beautiful white, scented flowers. At the time of writing, this species is known from 57 populations throughout seven states in the Great Lakes region. It requires undisturbed prairie habitat, which is now very rare, and the orchid is classed as threatened by the US government. Conserving the orchid involves tackling the variety of threats faced by the species as well as finding safe havens both in natural areas and botanic gardens. Active management of the habitat is required to remove aggressive invasive species and vigilance is needed to ensure that harvesting from the wild does not take place. In the past, over-collection by orchid growers has been a particular threat to this lovely species. Another threat is the loss of the orchid's pollinator. The nocturnal hawkmoth that pollinates the plant is considered to be an agricultural pest, and is being eliminated by insecticides.

A group of dedicated volunteers now uses hand-pollination to produce seed for use in the recovery plan. Chicago Botanic Garden helped them determine the most appropriate number of flowers to pollinate and produce seeds without over-taxing the plant and potentially leading to an early death. Climate change poses a new threat; as the local climate dries and warms, it is possible that the orchid may be unable to adapt or migrate, and it may become extinct. Garden researchers have looked at the genetic composition of this species, and a mapping project will help predict the future impacts of climate change on remaining populations. Saving the eastern prairie fringed orchid will require a sustained concerted effort

The attractive orchid *Platanthera leucophaea* is being studied and conserved by botanists at Chicago Botanic Garden. It may be necessary to assist its 'migration' to new sites as climate change affects its native habitat, which is already fragmented.

from the numerous federal agencies, land managers, researchers and volunteers involved in its conservation.

Researchers at Chicago Botanic Garden are also concerned about the future of Pitcher's thistle (*Cirsium pitcherii*) in the face of climate change. Native to the dunes along Lake Michigan, this thistle disappeared from the state of Illinois around 1915, largely due to of the enchroachment of urban areas on its lakeside habitat. The species is now classified as threatened throughout the US. It is a distinctive plant, growing 60 cm tall with a cream-coloured flower head and fuzzy white leaves. To restore Pitcher's thistle to Illinois, Chicago Botanic Garden has worked with science staff at the Morton Arboretum for almost 15 years.

Today, a reproducing population of this species is once again found on the dunes of Illinois, but it requires continual monitoring and supplementary planting from *ex situ* stock to ensure its long-term health. Unfortunately, predictive models created by garden staff indicate that the climate in Illinois where the plants now grow will no longer be suitable for the species by 2100. Again, the survival of this species will need continued collaboration between a wide range of agencies. Garden staff continue to help with monitoring and work on the genetics of the species, while the Holden Arboretum acts as the CPC National Collections Custodian for the plant, and many organizations work to enhance its chances of survival.

Another species that Chicago Botanic Garden is helping to conserve is the prairie bush clover (*Lespedeza leptostachya*), a rare plant that grows up to a metre tall and is very slender with pale pink flowers. The prairie bush clover now survives in only a few locations within a narrow range across the Midwest. Scientists believe that the disturbance historically caused by bison allowed this clover to survive among the dense vegetation that makes up its grassland habitat. Some of the last remaining populations of prairie bush clover in Illinois are found in a former agricultural area known as Nachusa Grasslands.

Researchers are now investigating whether disturbance by cows might fulfil the same ecological role that the bison once did, and are testing whether this endangered plant benefits from cattle grazing, which results in more light, nutrients and bare earth. Conservation scientist Dr Pati Vitt, who heads the project for the garden, explains that natural fires together with grazing and trampling by bison were always part of prairie ecology: 'We think that *Lespedeza* takes advantage of these disturbances to regenerate,' she says. 'If we can understand this process better we will be better able to manage the species and its habitats.' The research study involves bringing cows back to Nachusa Grasslands in a strictly controlled way. Within 90 test plots, some grazed and

Pitcher's thistle. US botanic gardens and conservation agencies are working together to ensure its survival.

some not, researchers monitor the coverage and abundance of the clover as well as the number of new plants. To supplement the fieldwork, new methods are being used to enhance seed germination and determine the genetic diversity of the Nachusa population so that plants can be produced for potential reintroduction. Meanwhile, seeds are kept safe in cold storage in the garden's seed bank. Havens is convinced that the efforts are worthwhile. 'It has been extremely rewarding to watch the conservation progress for this species. Our role at Chicago Botanic Garden has been to contribute to both the science and the application of the science. We also ensure long-term storage of plant material in our seed bank because we are the primary custodian for this species under the CPC's National Collection of Endangered Species scheme. *Ex situ* conservation is an important complement to *in situ* conservation because I am not convinced that the current natural habitat for this sensitive species will remain suitable with climate change. We need to prepare for very uncertain ecological conditions.'

the most
effective plant
conservation is
often experimental

## REFINING CONSERVATION PRACTICE

An important part of the work to recover populations of threatened plant species involves testing and improving *ex situ* conservation methods. The Chicago Botanic Garden has a seed biology laboratory that serves as a regional resource for seed cleaning, storage, viability testing and determining germination protocols. Genetic studies assess the implications of making single or multiple seed collections in a given year, and scientists also investigate the loss of genetic variation when seeds are stored. In 2003, the Dixon National Tallgrass Prairie Seed Bank was established, and it subsequently joined 'Seeds of Success', a consortium of US conservation organizations and agencies collaboratively collecting seeds across the United States. Seeds collected by the consortium are also deposited at the Millennium Seed Bank at the Royal Botanic Gardens, Kew.

The Dixon National Tallgrass Prairie Seed Bank is committed to collecting and banking seeds of the entire tallgrass prairie flora, approximately 1,500 species. Currently, the seed bank houses over 950 accessions of nearly 700 species, strategically chosen from across the region with priority given to species with high restoration value. Seeds are gathered by over 35 associate collectors, from populations in more than 45 natural areas across 16 states containing native prairies. Seeds are dried to less than

15 per cent humidity in a commercial seed drier, then cleaned to remove chaff, counted and/or weighed to estimate how many have been collected, and finally placed into the collection in a -20 degree freezer. Over 1,200 volunteer hours were donated to the seed bank in 2008, primarily by groups who assisted with seed processing.

Genetic studies are particularly important in the conservation of American beachgrass, (*Ammophila breviligulata*), a species that is under threat in the Great Lakes area and is being used quite extensively for habitat restoration. An important goal of native plant restorations is to reconstitute populations that are genetically similar to native ones, because this makes it more likely that the new plants will successfully establish and survive. Research led by Dr Jeremie Fant at Chicago Botanic Garden has shown that American beachgrass planted from commercial sources was of limited genetic variability and not of local provenance. The study underscores the importance of understanding the genetics of remnant native populations and restoration propagules before starting restoration efforts, especially when the populations are threatened or endangered.

The garden's Daniel F. and Ada L. Rice Plant Conservation Science Center has been certified by the Leadership in Energy and Environmental Design (LEED) green building rating system. It provides state-of-the-art laboratories and teaching facilities that allow garden scientists to continue pioneering new techniques for plant conservation. With a viewing gallery and green roof open to the public, the centre provides an important platform for conservation education and outreach to visitors and school groups, prompting visitors into further action.

Members of the public can become involved in Project BudBurst, a programme that uses citizen scientists to track plants through the seasons across the US and provides educational resources on climate change and phenology. The garden has also developed the Floral Report Card programme, a network of identical gardens planted in several US locations. These gardens, monitored by volunteers and students, are used to compare responses of cloned plants to different climates and provide valuable data about the effects of climate change.

## TRAINING THE NEXT GENERATION

Chicago has linked up with Northwestern University to offer a unique doctoral and masters programme in plant biology and conservation. The garden also partners the University of Illinois at Chicago on the LEAP (Landscape, Ecological and Anthropogenic Processes) doctoral programme, focusing on the ecology of landscapes altered by humans. Andrea Kramer, BGCI (US) Executive Director, based in Chigaco, believes that these have broad applications well beyond the Midwest: 'When faced with a worldwide plant extinction crisis we all benefit from sharing our experiences and techniques. Chicago is leading the way with genetic research and ecological restoration and is always ready to lend a hand to gardens facing problems with species conservation methodology in other countries. The most effective plant conservation is often experimental – each species having different characteristics and requirements. Whereas not all gardens have the technical resources enjoyed by Chicago many more could involve visitors and citizen scientist volunteers in the pioneering way developed by the garden.'

The Daniel F. and Ada L. Rice Plant Conservation Science Center. Conservation scientists have access to state-of-the-art laboratories and teaching facilities and members of the public can visit a viewing gallery to watch scientists at work.

The plants of Hawaii share the fate of all tropical island floras – the threat of imminent extinction. Lying over 2,000 miles from the nearest continental land mass, the islands of Hawaii are the most isolated high islands in the world. Plants arrived there as the spores and seeds of 'founder species' carried by the wind and sea and are now adapted to survive in a wide variety of different habitats. The flora is unique, with more than 90 per cent of about 1,200 native plants growing nowhere else in the wild. Around half of these naturally rare species are threatened with extinction as a result of deforestation, introduced insects and diseases, fire, grazing by introduced livestock, competition with invasive plants and climate change. As temperatures increase species with limited natural distributions may have nowhere to go.

**Previous page:** Sunset over the Kalalau Valley, on the Hawaiian island of Kauai.

The natural vegetation of Hawaii has been severely modified and now only a few remnants of forest remain on the mountain slopes. In the wetter areas giant tree ferns, lobelias, and a wide variety of other native plant species are found in rainforest dominated by the tree ohia lehua (*Metrosideros* spp.). The koa (*Acacia koa*) is dominant in dryer areas of woodland. Above the cloud line, on the peaks of Maui and Hawaii, the largest islands, are patches of more open woodland with low scrubby plants and scattered grasses. On the highest peaks there is an alpine vegetation zone with mosses and lichens. Patches of wetland are important havens for endemic plants and birds, such as the Hawaiian duck (*Anas wyvilliana*) and Hawaiian coot (*Fulica alai*). The dry coastal lowlands and the lower mountain slopes on the leeward side of the islands are now dominated by alien plants, such as algaroba (*Prosopis*, also known as mesquite), cacti and drought resistant grasses. Only occasionally are endemic species, such as the wiliwili tree (*Erythrina sandwicensis*) encountered in these damaged lowland ecosystems. The wiliwili is under further threat of extinction from an alien invader, the erythrina gall wasp (*Quadrastichus erythrinae*).

## IMPACT OF HUMAN COLONIZATION

It is still unclear when humans first found Hawaii but it is thought that Polynesians from the Marquesas travelled thousands of miles by canoe and settled around 1,000 years ago. The Polynesians brought with them food plants, such as the coconut, now common along Hawaii's beaches, the mountain or Malay apple with its beautiful red fruits, breadfruit, yam, banana, sugar cane, arrowroot, and taro.

According to island mythology, the first Hawaiian was the younger sibling of the first taro plant, which emerged after the burial of the stillborn son of two gods. Taro has great cultural significance in Hawaii and the tubers are eaten like potatoes or mashed, mixed with water and slightly fermented to make poi, a staple local food. Polynesian settlers also brought with them the candlenut tree, or kukui (*Aleurites mollucana*), which is now the Hawaiian state tree.

The rats, chickens and small pigs brought by the Polynesians disrupted the nests of giant flightless waterfowl, at least five species of which were hunted to extinction. Clearance of native vegetation for farming started the process of plant extinctions which continues today. The impact of the Polynesians on

Taro (*Colocasia esculenta*) growing at the National Tropical Botanical Garden, Hawaii.

most of the
ornamental
plants associated
with Hawaii are
introduced
species that
have replaced
the delicate
native flora

Hawaii's biodiversity was significant but not on the same scale as the devastating impact of later settlers. Real ecological devastation followed the third Pacific expedition of the British explorer, Captain James Cook. His arrival in 1778 heralded a new phase for the remote islands, opening up Hawaii to the outside world and beginning a period of rapid social and ecological change. This coincided with the rise in power of the local ruler, Kamehameha the Great. By 1810 Kamehameha had unified the main islands for the first time.

A major natural commodity that first caught the attention of European and American traders was sandalwood (*Santalum* spp.). The demand for this was huge –mainly from China. The sandalwood trade was disastrous both for the farming communities of Hawaii and for the native forests. Men were forced by Kamehameha to cut sandalwood first in the lowlands and then in the mountains and transport the logs on their backs to the sea. The export of sandalwood paid for Kamehameha's acquisition of European weapons, warships and luxury goods. The selective removal of sandalwood was quite wasteful, and many other trees were cleared in the process. This opened up the forests to the invasive plants and animals that have wrought such destruction on Hawaii's ecosystems. Now only a few sandalwood trees remain.

Along with cattle, sheep and goats, Europeans and Americans settling in Hawaii introduced numerous kinds of exotic flowering plants. These include pines and mesquite from North America, the silky oak (*Grevillea robusta*) and eucalyptus from Australia, frangipani (*Plumeria*) and guava from tropical America, Bermuda grass from southern Europe, and gorse from western Europe. Most of the ornamental orchids, hibiscus, gingers, jacarandas, and poinsettias associated with Hawaii are also introduced species that have replaced the delicate native flora of these tropical islands.

Plants have been introduced for food, forestry, ornamental display and – perhaps most misguided of all – for revegetation. By the middle of the 19th century, forest destruction above Hawaii's capital Honolulu, on the island of Oahu, had altered the local climate to such an extent that there were worries about the city's water supply. The response to these concerns was to reforest the area with fast-growing imported species. As recently as 50 years ago non-native woody plants were intentionally spread by aeroplane in an attempt to 'improve' the native forest.

## EXPLOITATION AND EXTINCTION

The impact of exploitation, clearance of vegetation and competition with introduced species has led to the extinction of many native plant species. Already 55 of Hawaii's plant species are recorded as extinct and a further 42 probably no longer exist in the wild. Another 150 species are reduced to less than 50 individuals and need intense intervention to ensure their survival.

Their future is uncertain but botanists at the National Tropical Botanical Garden, headquartered on the island of Kauai, are determined to rescue native plant species from extinction. The botanic garden was chartered by the US government in 1964 as a privately supported research and education institution. It now encompasses more than 1,600 acres across four gardens and three nature reserves in Hawaii and also has a garden in Florida.

Restoring Hawaii's coastal and lowland habitat is a major goal of the National Tropical Botanical Garden.

The conservation work and adventures of botanists Steve Perlman and Ken Wood are legendary. 'As long as one individual remains in the wild we can save a plant species from extinction,' says Perlman. 'We don't give up until there is absolutely no chance.' Perlman and Wood routinely witness the processes of extinction at first hand and quite probably hold the record for saving more species than any other conservationists in the world. As well as monitoring and collecting conservation material from the most endangered plants over the past 20 years, they have helped to find and relocate roughly two dozen species that were previously thought to be extinct.

Concentrated conservation efforts are now a top priority for the critically endangered Hawaiian plant species that have fewer than 50 individuals remaining in the wild. These species are included in Hawaii's Plant Extinction Prevention Programme. Perlman and Wood follow strict protocols as they locate and rescue some of the world's rarest plants. Their knowledge of where these species cling onto survival has been built up over several decades of fieldwork and exploration of the remotest and most inaccessible locations by helicopter, kayak or climbing rope. Many endangered species in Hawaii can now only be found in the most remote locations, the steep slopes of volcanoes, deep rocky ravines and offshore rocks that are pounded by waves – some of the few places where plants are safe from browsing goats and pigs.

Working as a team, Steve Perlman, Ken Wood and their young local protégé Natalia Tangalin plan their work in consultation with Dr David Burney, Director of Conservation at the National Tropical Botanical Garden. In the field, they follow strict guidelines to ensure that a representative sample of genetic diversity is collected – to the extent possible with so few plants remaining. They keep meticulous records of locations of individual plants and populations so that high-quality species distribution and survey maps can be created and when the seeds or cuttings are brought back to the garden accession details for the precious wild material are carefully recorded. Young plants are established in cultivation and carefully nurtured, with records kept of actions taken throughout the process. When possible, seed is stored to ensure the long-term *ex situ* conservation of severely threatened species. The goal is to provide a temporary safe haven until plants can be safely reintroduced to the wild.

## LOULU

The garden's track record in raising the endangered plants of Hawaii is demonstrated by the presence of well over 100 endangered species in its collections. Among these are the endemic loulu or *Pritchardia* palms. Named after William T. Pritchard, a 19th-century British consul in Fiji, *Pritchardia* is a genus of about 27 species found in Hawaii and other islands of the Pacific. Twenty-two species are found in the Hawaiian archipelago, each confined to a single island. The  palms face all the familiar Hawaiian threats of grazing animals, introduced insects, invasive weeds competing with native seedlings and general habitat degradation. In addition they suffer from illegal collecting of seeds and young plants by people who want to add these rare palms to their garden collections. Only two Hawaiian species survive as reproducing wild populations, *Pritchardia remota* and *Pritchardia hillebrandii*. The very rarest species are *Pritchardia aylmer-robinsonii*, with only two trees left on the island of Niihau and *Pritchardia munroi*, reduced to one individual on the island of

Molokai. The National Tropical Botanical Garden aims to have all the Hawaiian species in its conservation collection with up to 50 trees per species, where possible, to ensure genetic representation of surviving populations. Some of the palms in cultivation at the garden are now over 25 years old.

Garden staff are actively researching and monitoring fragile *Pritchardia* populations. The wild plants are protected from rats by applying poison and guards to stop predation of the fruits and increase the chances of natural regeneration. Wild populations are also given a boost with with cultivated palms. In some of the loulu restoration sites, just one species (such as *P. aylmer-robinsonii*) will be used to create new populations; in others two species that naturally occur within the same range (for example *P. viscosa* and *P. hardyi*) will be planted out in the same area. *Pritchardia napaliensis* and *P. limahuliensis* have been planted at the Limahuli Preserve managed by National Tropical Botanic Garden. The upper valley of Limahuli is mostly intact and is home to an exceptionally high diversity of native plants and animals. Among these are populations of ten plant species classed as threatened or endangered under US

Limahuli Garden on Kauai's north shore, part of the National Tropical Botanical Garden system, features centuries-old terraces for growing taro, the traditional Polynesian staple. This spectacular ancient Hawaiian garden is part of a 'mountains-to-sea' management programme that includes the 1000-acre Limahuli Preserve. Following ancient principles of 'ahu`pua`a' land management, garden staff participate in integrated conservation projects from the coral reefs offshore to the summits of the jagged peaks beyond.

national conservation legislation, a large nesting colony of two endangered seabirds (Newell's shearwater and Hawaiian petrel), Hawaiian honeycreepers, Hawaiian owls and the endemic hoary bat. To help ensure the long-term protection of this crucial site a fence has been constructed which encloses the entire upper valley of Limahuli, designed to keep feral pigs and goats out of the reserve.

Fewer than 100 individuals of the Critically Endangered *Pritchardia limahuliensis* are known from Limahuli Valley, where its natural habitat is lowland moist forest. Regeneration is limited, mainly because of seed predation by rats and pigs. Since 1998, more than 10,000 nursery-grown native trees and shrubs have been planted across over 6 hectares of specially designated sites in the lowland forest of the Limahuli Preserve. Along with the endemic palm, other plants being used in this restoration effort include the Kauai endemics *Munroidendron racemosum* and *Kokia kauaensis*.

*Pritchardia limahuliensis*, a Critically Endangered palm species that survives with the help of the National Tropical Botanic Garden.

## HIBISCUS

Hibiscus flowers have become a popular symbol of Hawaiian culture, often used in the traditional leis or flower garlands in a custom introduced to Hawaii by the Polynesian settlers. Now more than 5,000 varieties of hibiscus are grown on the islands but the native *Hibiscus* and its relations in the family Malvaceae include some of Hawaii's most threatened plants. One such species is the Critically Endangered *Hibiscus clayi*, a small tree with brilliant red flowers which is reduced to just four individuals in the degraded montane forest of eastern Kauai. Cattle and feral pigs have contributed to the decline of this species but fortunately this is one of the National Tropical Botanic Garden's conservation success stories.

In early efforts with this species, Burney and his colleagues had found that one could clone the four remaining individuals easily, but they rarely produced viable seed. Then, thanks to David Orr of the Waimea Arboretum on Oahu, who is an avid collector of Malvaceae, Burney learned that there were four

# KOKIA COOKEI — BACK FROM THE DEAD?

The genus *Kokia* comprises four species, all found only in Hawaii and all classified as Extinct or Critically Endangered. *Kokia cookei* is now extinct in the wild. When first discovered in the 1860s only three small trees of the species were found growing in dryland forest at an elevation of around 650 ft in western Molokai. The site formed part of a sheep run and the plant population was directly affected by being browsed and trampled by domestic and feral animals. Other threats that contributed to the species' decline included habitat conversion, loss of native pollinators, and seed predation by insect larvae. By 1918 the wild plants had all died. Some years later, the only known cultivated tree died without producing viable offspring and the species was thought to be extinct. However a cultivated specimen was discovered in 1970 and living material from a branch has been grafted onto the more common *K. kauaiensis*. Currently, *Kokia cookei* exists in cultivation at two locations and in managed outplantings at three sites. Private individuals and non-governmental organizations have made efforts to increase the number of plants of this species – using seeds, cuttings, grafting, tissue culture and air layering – and a recovery plan has been developed under the US Endangered Species Act. The conservation of this plant is not without controversy, however. Local landowners and the US government conservation agency are in conflict about the designation of land for conservation and the role of private landowners in the propagation and management of the species. Whatever the outcome of this debate, *ex situ* conservation skills will be essential to ensure the plant's survival.

other genetic strains still thriving, as a result of collections made by legendary botanist Joseph Rock and taken to Kew. Cuttings from these four were obtained, and new populations based on these eight individuals began producing copious seed in their restoration sites within a few months.

Sometimes it is too late to save the natural populations of a plant species in the precise locality where it is found. This happened with one of the native relatives of *Hibiscus*, a small tree with a grandiose name: *Hibiscadelphus hualalaiensis*. The last individual of this species surviving in the wild was fenced to protect it from grazing cattle but it nevertheless died in 1992; the species is now extinct in the wild. However, Steve Perlman collected seed capsules from the dead plant and seeds were sown at the National Tropical Botanical Garden. The resulting cultivated plants have been planted out and the species still survives close to its native location today.

**Opposite:** *Hibiscus kokio* subsp. *saintjohnianus* is a rare plant growing only in the north-western part of Kauai.

## BRIGHAMIA

*Brighamia insignis* is an unusual plant that belongs to an endemic Hawaiian genus of only two species. Native to the islands of Kauai and Niihau this plant is now well established in the horticultural trade but only a single plant remains in the wild. An extraordinary member of the bellflower family, *Brighamia* is commonly known as Hawaiian palm and sometimes called 'cabbage-on-a-stick' because of it's strange form. A rosette of somewhat shiny leaves grows atop a succulent stem, which usually reaches a height of between 1 and 2 metres but can grow as tall as 5 metres. The fragrant creamy yellow blooms are trumpet shaped with a long floral tube. The shape, colour and scent of the flower suggests that it was pollinated by a long-tongued moth, and botanists suspect that this may have contributed to the Hawaiian palm's downfall. Moth species of this right size are extremely rare, possibly even extinct, on Kauai and without a pollinator the species is likely to be doomed in the wild.

Steve Perlman and Ken Wood have spent many years recording and collecting the species from its last strongholds – Hawaii's sea cliffs. 'We've watched natural populations of this plant plummet in the wild, but working by rope we managed to pollinate wild plants on the sheer cliffs, collect seed, establish *Brighamia* in cultivation, and then work with commercial nurseries. Now this plant is being sold all over the world. A success story!' One of the places *Brighamia insignis* grows in splendour is the US National Botanic Garden close to the Capitol Building in Washington DC. A permanent display of Hawaii's endangered plants is grown close to this seat of power – a reminder to the thousands of visitors of the fragility of Hawaii's flora.

The alula, *Brighamia insignis*, is a beautiful Hawaiian endemic in the bellflower family that, although down to just one individual in its original habitat, has thrived in restoration sites on Kauai and as a popular house plant as far away as Europe.

## KANALOA

*Kanaloa kahoolawensis*, known as Ka palupalu o Kanaloa or kanaloa, is an endangered species discovered in 1994 on a sea stack near a former bombing range on the island of Kahoolawe. That individual is the only known example of the species surviving in the wild; one other plant, grown from seed, is found in the McBryde Garden on Kauai. In contrast to today's situation, the pollen record indicates that before human arrival kanaloa was widespread and common in the

lowlands of Oahu, Maui and Kauai. Wood suspects that the plant fell victim to habitat destruction and exotic grazing animals, and to an indirect ecological effect. Since rainforest on nearby Maui was destroyed, the air carries less moisture from plant transpiration to Kahoolawe, causing drought conditions.

The nursery of the National Tropical Botanical Garden is trying to propagate the two surving plants in order to establish *ex situ* stocks and new populations. The goal is to establish three kanaloa populations of at least 25 individual plants within the historic range of the species that will reproduce naturally and increase in number. This is proving to be a very challenging goal. Although the two plants occasionally produce viable seed, few seedlings survive, and the roots seem to be susceptible to fungal attack. Nevertheless, hope remains for a few surviving small plants being cared for in three collections. Further research is needed in order to establish which species pollinate the plants and to determine what conditions the seedlings need to survive.

## THE FUTURE

Faced with urgent need to conserve so many plants, Burney is a pragmatist: 'We just have to do what we practically can to save, restore and protect as many species as possible. We need to convince people of the need to act now. We cannot go back in time but we can learn a lot from the past.' He is developing a new approach to plant conservation in Hawaii. This involves looking at ancient records, such as fossil remains, to develop restoration plans for badly degraded areas. Understanding the past distributions of native plants allows species to be reintroduced to sites where they have not been growing wild for decades or even centuries. This helps to solve a 'storage' problem for endangered species that have been taken into *ex situ* conservation collections and successfully propagated. Burney considers this form of conservation to be *inter situ*.

'This really means reintroducing species to locations outside their current tiny discontinuous range to where they grew in the past – largely before human interference. As part of the restoration process we can substitute living relatives or ecological surrogates for globally extinct species that had an important ecosystem role in the past. At one of our sites, we have successfully established 77 different plant taxa on a three-hectare plot of abandoned farmland. In essence we are beginning to re-wild Hawaii.'

The Lawai Kai Coastal Restoration site is located at the mouth of Lawai Stream, on the south shore of Kauai adjacent to the beautiful MacBryde Garden.

Field botanists from the National Tropical Botanical Garden collect seeds for restoration projects.

It is near palaeoecological and archaeological sites, which provide information on the local ecological history that has guided the restoration plan. The restoration project aims to improve coastal and lowland forest habitat for more than 20 rare native plant species and to remove a thick mat of alien grass from the beach strand to enable sea turtles to nest.

At Makauwahi Cave Reserve, leased from the Grove Farm Company, the garden's large neighbor on Kauai's south shore, the richest fossil site in the Hawaiian Islands has provided a very detailed picture of what lowland Kauai was once like. Using this information, Burney and his wife, Lida Pigott Burney, have created plant restorations based on the indications from fossil pollen, spores, seeds, and wood. Volunteers from the community and visitors from throughout the world are welcome to tour the site, assist with the restorations, and carry out research on the six types of 'inter situ' restoration projects there. A summer field school programme brings students in archaeology, paleo-ecology, natural history, ethnobotany, and botanical illustration to work and study at the site.

Every restoration and protected site in Hawaii is a precious biodiversity resource. The conservation challenges are immense, not least in convincing people of the importance of Hawaii's unique flora. This can be more difficult that one might imagine, as the goals of plant conservation may contradict other land uses in areas receiving some measure of supposed 'protection'.

The local hunting lobby has a strong political influence in Hawaii, and likes to see healthy populations of the feral pigs, goats, and deer that continue to cause such devastation in the few areas of near-natural vegetation. At the same time, many rare species in Hawaii are being grown *ex situ* and reintroduced with only skimpy knowledge of their biology and ecology. The National Tropical Botanical Garden has been a leader in promoting research on all aspects of endangered plants, from plant taxonomy and biogeography to basic ecology and horticulture.

Burney sums up the challenge: 'The only hope for these hundreds of declining species is to try everything we can think of, and monitor the results – and do it now, while there are still plants to work with.'

we just need to do what we practically can to save, restore and protect as many species as possible; we need to convince people to act now

Mexico is the centre of diversity of cacti with around 430 endemic species, ranging from the bat-pollinated, tree-like columnar cacti to the tiny living-rock cacti that were so valuable in the 19th century. For a short time rare species of cacti, such as *Ariocarpus kotschoubeyanus* were worth more than their weight in gold. Now cacti are grown commercially on an enormous scale and are widely available in supermarkets and garden centres. Nevertheless, specialist plant collectors around the world still crave wild specimens of newly described rare species and collecting pressures are one of the main threats faced by cacti in the wild. Around one third of all cacti are threatened with extinction in their natural habitats. Small, bizarrely shaped plants have been especially attractive to cactus enthusiasts overseas, particularly as they can be challenging to grow.

**Previous page:** The magnificent *Agave americana*, seen here in Mexican oak woodlands, is grown in many botanic gardens and is cultivated widely as an ornamental. It has become naturalized in many parts of the world. Other species of *Agave* are threatened with extinction as a result of habitat loss and unsustainable use.

Species of *Ariocarpus*, *Astrophytum*, *Aztekium*, *Obregonia* and *Pelecyphora* are among those that have been collected close to extinction for the international market. *Aztekium hintonii* was first discovered in 1991 in the mountains of the Sierra Madre Oriental at the same locality as *Geohintonia mexicana*, a new genus comprising a single species described in 1992. Both these species grow on gypsum cliffs and hillsides in a remote area that is not currently farmed or subject to any development. The main threat to both species has been illegal collecting. Shortly after the species were described in the scientific literature both were available for sale in European nurseries – contravening Mexican and international legislation.

Mexico also has a rich diversity of other succulent species, including 140 species of agave and 23 species of yucca. The agaves are important in the rural economy of Mexico as a source of fibres used to make mats, baskets and paper and for the production of traditional alcoholic beverages. *Agave tequilana* is cultivated commercially for the production of tequila and other agave species are harvested from the wild to produce mescal and pulque.

Dr Héctor M. Hernández has long been involved in the study and conservation of Mexican cacti and their natural habitats. He now coordinates efforts to conserve cacti and other succulent plants worldwide as Chair of the IUCN/SSC Cactus and Succulent Specialist Group. He believes that accurate information is important as a basis for planning effective conservation action. 'Even now some parts of Mexico are poorly known from a botanical point of view and we have a lot still to learn about the distribution and ecology of our native cacti and succulents. Nevertheless, we know enough to conserve some of the more important sites for cactus and succulent diversity and we need to back this up with well-documented *ex situ* plant collections for our rarest species.'

## UNAM BOTANIC GARDEN

The botanic garden of Universidad Nacional Autónoma de México (UNAM), situated on an ancient lava flow 2,250 metres above sea level in

**Left:** *Ariocarpus kotschoubeyanus* is a Mexican endemic cactus quite widely distributed with scattered populations from the state of Coahuila in the north to Queretaro in the south.

**Right:** *Astrophytum myriostigma.* This yellow-flowered succulent occurs in northern and central parts of Mexico and is a popular plant in cultivation.

**Mexico has banned the export of wild-collected cacti for over 60 years but cactus enthusiasts still take plants from the wild**

Mexico City, has one of the most impressive collections of cacti and succulents in Mexico. A major portion of the garden's 12.6 hectares is devoted to succulent species, including Mexico's National Agave Collection which has over 140 of the 251 Mexican agaves with many magnificent large specimens. The national collection of Crassulaceae is also found at the garden and includes 236 out of the 376 species of *Echevaria* known from Mexico. A considerable proportion of Mexican cacti is also represented by the 381 species in the collection, including many that are rare and threatened.

Mexico has banned the export of wild-collected cacti for over 60 years but this has not always deterred cactus enthusiasts visiting from overseas. The entire cactus family is covered by CITES, which encourages international collaboration to prevent cactus smuggling. Nevertheless the digging up of wild plants continues. The UNAM Botanic Garden acts as a rescue centre for illegally collected, confiscated plants and some species brought into cultivation when their habitats were destroyed. UNAM helped to secure plants of *Echincactus grusonii*, for example, when their natural habitat was destroyed by flooding after dam construction.

The Director of the garden, Dr Javier Caballero, considers the emphasis on succulent plants to be extremely important. 'Mexico is renowned worldwide as a centre of plant diversity and our unique richness of succulents is an important part of our natural heritage. UNAM acts with the Mexican Association of Botanic Gardens to ensure that all species are represented in well documented living collections within the country.'

The living plant collections are important for teaching and of considerable value for taxonomic and evolutionary research. An active programme of propagation is carried out for endemic species and plants are made available to local schools to promote an interest in Mexico's unique flora. Propagated plants are sold to the public to discourage the sale of plants collected illegally from the wild.

Cultivated material also has the potential to restore populations in the wild. This has happenedwith a local form of *Mammillaria* known as *san-angelensis* that was reintroduced to a small protected area (Reserva Ecológica del Pedregal de San Ángel) within the main University campus in Mexico City.

The UNAM Botanic Garden has a wonderful collection of cacti and other succulents. The plants are used for display, education, research and conservation.

Propagation of endangered cacti is an important component of their long-term conservation. Over half the globally threatened species of cacti are in cultivation at botanic gardens

## THE CHIHUAHUAN DESERT

The work of Héctor Hernández has contributed directly to the conservation of cacti both in the wild and in *ex situ* collections. One of his projects has involved mapping the general pattern of distribution of threatened cacti within the Chihuahuan Desert – one of the three semi-desert areas of Mexico that are most important for cactus diversity and conservation. Within the Chihuahuan Desert the richest area is around the boundaries of the Nuevo Leon, Tamaulipas and San Luis Potosi states. Hernández and his colleagues used herbarium data from the National Herbarium of Mexico and 34 other institutional herbaria around the world to develop a distribution database for cacti to use for GIS mapping and analysis. Based on the results a recommendation for a new protected area was made to the State Government of San Luis Potosí and in 1997 the Guadalcázar Protected Natural Area was established. One of the specific objectives of the reserve is to conserve the extraordinary wealth (76 species) of cacti in the area, in particular the rare and endemic species such as *Ariocarpus bravoanus, Coryphantha pulleineana, C. odorata, Mammillaria aureilanata, M. microthele* and *Turbinicarpus knuthianus,* among many others.

One significant species discovered by Héctor Hernández is not yet in cultivation in the UNAM Botanic Garden. In the wild *Ariocarpus bravoanus* is known only from a small area at the edge of the Chihuahuan Desert within the state of San Luis Potosí. This is the most recently described species of the genus of living-rock cacti and is considered to be endangered mainly because of the continuing threat of collectors.

## THE GOLDEN BARREL CACTUS

The distinctive golden barrel cactus, now a common plant in botanic garden displays around the world.

*Echinocactus grusonii* was first discovered in 1889 and immediately an international demand was created for this distinctive cactus, prized more for its handsome growth form and golden spines than its relatively small yellow flowers. Plants were exported around the world and by the end of the century concerns were expressed about the potential extinction of the species. In the wild the golden barrel cactus has a limited distribution in the Moctezuma River Canyon between the states of Querétaro and Hidalgo. In 1995 a dam was built across the river to provide hydroelectric power, flooding most of the habitat of *Echinocactus grusonii*. Then in 2005 a population was unexpectedly discovered about 500 km away from the original locality, in the state of Zacatecas. The DNA structure of the plants from the new and original locations is currently being studied by researchers from the Universidad Autónoma de Querétaro, the University of Reading, and UNAM in order to develop a certification scheme. The results from this investigation will make it possible to trace the geographical origin of particular plants by means of molecular markers. *Echinocactus grusonii* is considered to be Critically Endangered in the wild but is widely cultivated: over 100 botanic gardens have this species in their collections. For *ex situ* conservation purposes plants of known wild origin are particularly valuable as they form part of the original gene pool of the species.

# DAHLIA — A NATIONAL TREASURE

Cacti are commonly associated with Mexico but other very significant garden plants also have their origin in this botanically rich country. Dahlias are one such group. Named after Anders Dahl, a pupil of Linnaeus, dahlia species were introduced to Europe in the early 19th century and a huge range of cultivars were developed.

In their native Mexico and neighbouring countries dahlias were once used as a medicinal plant and the tubers were eaten as a vegetable. Many ornamental varieties were also produced in the ancient gardens of Mexico and the dahlia is Mexico's national plant. UNAM Botanic Garden has an important collection of dahlias including the species *Dahlia brevis*, which is now extinct in the wild.

Hugely popular as garden plants around the world, dahlias can be seen growing wild in Mexico and Central America.

### TEHUACAN—CUICATLÁN BIOSPHERE RESERVE

Another of the most important areas for cactus conservation in Mexico is the Tehuacan-Cuicatlán Biosphere Reserve, an area that has been studied in detail by botanists from UNAM. The reserve, in the States of Oaxaca and Puebla, has almost 3,000 vascular plant species in an area of 10,000 km2. There are dense wooded sites of tree cacti along with a range of agaves and other succulents such as *Fouquieria purpusii* and *Beaucarnea gracilis*. UNAM provided a safe haven for plants of about 50 cacti species of the area when a highway was constructed. More recently, an important project has been undertaken to assess the impact of land use and climate change on the threatened cacti within the Biosphere Reserve. Records of 82 cactus species growing in the reserve, including 21 that grow nowhere else, have been analysed to help scientists predict the likely effects of increases in temperature and lower rainfall. Predictions of the possible future distribution of the cacti of Tehuacan-Cuicatlán will be invaluable for planning long-term conservation strategies in response to global climate change. Results show that only a very few species are likely to extend their distribution area as climatic conditions alter. In fact, almost 95 per cent of species are expected to drastically reduce their areas and half are likely to be extinct by the end of the century. Species that are able to occupy habitats at higher altitudes are most likely to survive. Some vulnerable endemic species with small distributions, such as *Mammillaria huitzilopochtli, M. napina* and *M. pectinifera,* seem to be less sensitive to climate change than those with wider distribution, such as *Cephalocereus columna-trajani* and *Neobuxbaumia tetetzo.* Careful monitoring will be required to ensure that management practices for the threatened cacti take account of the changing climatic conditions.

Around the world many botanic gardens have collections of cacti and succulents. Most of the rare and threatened species are in cultivation so there is a tremendous opportunity to look at restoration efforts. This involves matching the plants of correct genetic provenance to the habitats were they originated. The successful work of Fairchild Botanic Garden with cactus restoration shows what can be done.

Brazil's oldest botanic garden, the Rio de Janeiro Botanic Garden was established in 1808 by Dom João VI, then Prince Regent of Portugal. The impressive avenues of Cuban imperial palms – an iconic landmark of the city – date from the garden's birth. Although founded to acclimatize new cash crops for the agricultural development of Brazil, in 1890 a decree specified that the garden should be dedicated to native plants as well as providing a site for recreation within the developing city, and a herbarium was established to provide a reference collection of the Brazilian flora. Today, the emphasis has shifted to conservation and ecological restoration. With the largest expanse of tropical rainforest on earth, Brazil faces huge conservation challenges. Botanic gardens have a special role to play in studying, safeguarding and restoring the country's valuable and threatened plant species.

**Previous page:** The Atlantic Forest of Brazil is a global biodiversity hotspot with a rich diversity of trees, birds and mammals. About 6,000 plant species are endemic to this forest ecosystem.

The largest area of Amazonian rainforest lies within the Brazil and Atlantic forest, another type of rainforest with a very different species composition, is a significant biodiversity hotspot. Deforestation of both types of forest continues apace with clearance for soya plantations, cattle ranching, urbanization, mining, road construction and illegal logging all contributing to the loss. As well as the rainforests Brazil has a vast area of cerrado vegetation characterized by savannah, woodland and dry forest. In the north-east, arid, scrubby caatinga vegetation covers 10 per cent of the country. Over 55,000 plant species are believed to be native to Brazil, considerably more than in any other country in the world, and botanical exploration is still incomplete.

Species from the Atlantic Forest, elsewhere in Brazil and beyond are all on display in the magnificent Rio de Janeiro Botanic Garden. Palms – one of the largest collections in the Americas – orchids, bromeliads and cacti are all well represented. In total, 1,700 tree species are included in the collections along with 190 palm species. Fabio Rubio Scarano, the garden's Director of Scientific Research, has written, 'An ark would be too small to confront the deluge indicated by the more drastic predictions of the globe's future climate and

environment. Thus the greatest capacity of the botanical garden is its capacity for anticipating the worst, through the generation of new knowledge applicable to practical and immediate solutions, and not to restraining the inevitable'.

## THE ATLANTIC FOREST BIOSPHERE RESERVE

The Rio Botanic Garden is designated as the nucleus of the reserve area of the Atlantic Forest Biosphere Reserve defined by UNESCO. The Atlantic Forest or Mata Atlântica is considered to be one of the world's global biodiversity hotspots. Over 20,000 plant species occur within this rainforest ecosystem, around 6,000 of which are endemic. About half the tree species are endemic to the Atlantic Forest, including valuable timbers such as the endangered Brazilian rosewoods, *Dalbergia nigra* and *Caesalpinia echinata*.

This rainforest, extending in a narrow belt for over 3,000 kilometres along the coast of Brazil, has been progressively deforested since Brazil's discovery by European settlers 500 years ago. Felling for timber, mining, and the development of agricultural plantations have all taken their toll. Now the region contains the most densely populated areas in Brazil, with over 100 million people, and only about six per cent of the original coastal forest remains. The Atlantic Forest Biosphere Reserve, established in 1992, includes important fragments of the Atlantic Forest in 14 Brazilian states. The main aim of the reserve is to conserve and restore ecological corridors as well as a significant proportion of the Atlantic Forest's biological diversity. A wide range of management, scientific and community organizations work together in an effort to manage this complex initiative, taking into account the need to develop sustainable use and social practices. The scientific, social and political challenges are major.

Tania Sampaio Pereira from the Rio Botanic Garden has stressed how important it is for botantists to work with all sectors of society if they are to make a difference in this region: 'We have to explain why it matters if the plants we love are destroyed. We must emphasize how each species is an important component of the forest contributing to the integrity of the ecosystem.'

## PAU BRASIL

One of the species the garden is helping to conserve is the rosewood *Caesalpinia echinata*, commonly known as pernambuco, brazilwood or pau brasil. This endangered tree species is an important flagship species for forest conservation; it gave Brazil its name and is unique to the Atlantic Forest. Pau brasil has been heavily traded for over 500 years, initially as a very valuable source of red dye in international trade. With the introduction of synthetic dyes in the 18th century the demand for dyewood collapsed but a new use was found for pau brasil.

Since the early 1800s the wood of pau brasil has been used for making bows for violins, violas, cellos and basses. Most professional bows are made from this timber, which is highly valued for its combination of durability, flexibility and resonance. No comparable substitute material has been found and despite conservation legislation in Brazil which bans the cutting and trade of pau brasil, international trade in the timber is estimated to be worth millions of US dollars a year and is likely to represent significant illegal exploitation.

Researchers at Rio Botanic Garden first became interested in the species in 1993 and remain deeply concerned about its future. Despite the huge cultural importance of pau brasil, remarkably little is known about its biology and the composition and structure of the plant community in which it occurs. The research team and indeed other groups in Brazil can propagate and grow it but fear that unless national and international legislation is rigorously enforced the few remaining wild populations will be lost. It will be necessary to persuade bow makers to accept plantation timber, which is unfortunately seen as being of inferior quality.

The garden's work on pau brasil over the years has involved sampling material from wild populations to undertake studies on the genetic make-up of the species. This will help in understanding the viability of the species and also whether certain populations may be genetically distinct. This knowledge will help determine the conservation management requirements of isolated populations, and might lead to useful information to use for selecting plants for plantation development. A team of experts looks at different aspects of the conservation biology and conservation of pau brasil and works with the people who live in the forest areas and rely on them for their livelihoods. The team believes that if they can convince local people to value and look after the species they may win the battle for its survival.

Most people interviewed in the communities of the Jacaré and Peró neighbourhoods (in the coastal municipality of Cabo Frio) and the José Gonçalves neighbourhood (Armação de Búzios) felt that conserving natural vegetation was the right approach while others considered the creation of more regional horticultural nurseries for local cultivation of the species to be very important. As part of the conservation effort local people have been paid to plant pau brasil seedlings in areas that already contained small populations of the species. People have also been encouraged to plant seedlings in public places and yards within the range of this valuable tree, to supplement the natural vegetation remnants and encourage gene flow between populations.

Pau brasil, the national tree of Brazil has been heavily logged for centuries for international trade. It was added to Appendix II of CITES in 2007.

## APPLIED RESEARCH

More broadly, Rio de Janeiro Botanic Garden is actively involved in the study and conservation of the remnants of the Atlantic Forest by undertaking research and practical measures that will help maintain protected areas and develop the vitally important connecting corridors. One of the garden's early successes was the creation of Brazil's first protected area, the Itatiaia National Park located in an area of high mountains on the border between the states of Minas Gerais and Rio de Janeiro. The Itatiaia Biology Station was set up in 1929 to facilitate botanical research in this unique area that was being progressively cleared for pasture and farming, and the national park was established in 1937. Botanists from the garden have since helped identify other important areas for protection within the Atlantic Forest and have participated in compiling inventories of protected areas to understand which species are present in the designated sites. Currently around one third of Brazil's Atlantic Forest is under some form of protection. Around 30 Private Natural Heritage Reserves have been established under a scheme where landowners guarantee protection of the animals and plants in the reserves in exchange for tax reductions.

Lack of knowledge about the germination requirements for tree species in the tropics remains a stumbling block in conservation and restoration efforts. Seed germination, storage and desiccation tolerance of tropical tree species are the main lines of research at the seed laboratory in Rio. One of the species that the laboratory has been working on is the palm *Euterpe edulis* – another important Brazilian species that is in trouble in the wild. Found in the Atlantic Forest and cerrado, the delicious edible palm hearts of this species are harvested extensively from the wild. Various efforts are underway to sustain wild harvesting as a means of providing livelihoods for local people and the scientific work at Rio Botanic Garden is supporting these efforts.

## CACTI

Cacti are usually associated with arid areas, but the Atlantic Forest has a surprising diversity of epiphytic species that hang from the branches of rainforest trees. Species that are found growing naturally on the trees in Rio Botanic Garden's arboretum include *Rhipsalis teres, R. pachyptera, R. lindbergiana* and *Epiphyllum phyllanthus*. *Rhipsalis pentaptera*, a common species in cultivation is now believed to be extinct in its native habitat, but the occasional specimen can still be spotted growing on old street trees not far from the Botanic Garden, in a now built-up area of Rio. Another species, *Rhipsalis mesembryanthemoides* is also heading for extinction in its natural rainforest habitat, but it can still be seen in the neighbouring Parque Lage. The dazzling *Schlumbergera truncata*, the Christmas cactus known in Brazil as flor-de-maio, is native to Rio de Janeiro and once occured naturally within the garden grounds. The garden celebrates mother's day each year with a colourful display of *Schlumbergera* cultivars, raising public awareness of horticulture and conservation.

we must explain why it matters that the plants we love are destroyed

## POÇO DAS ANTAS BIOLOGICAL RESERVE

One of the protected areas that has been surveyed in detail by botanists from Rio de Janeiro Botanic Garden is the Poço das Antas Biological Reserve, which was set up to protect the golden lion tamarin (*Leonthoptecus rosalia*). This internationally important reserve has patches of secondary forest surrounded by abandoned pastures with invasive exotic grasses. Frequent burning has hindered natural regeneration of the forest fragments and so the botanic garden has been helping with forest restoration. An initial step was to document the species that occur naturally in the regeneration sites. Following this, seed was collected for *ex situ* conservation purposes and methods for propagating the tree species were investigated. After trials with over 20 different species, trees that establish quickly and are able to prevent the invasive grasses from taking hold were selected for replanting.

After 16 years of work, the Poço das Antas experience is now providing a model for restoration of other degraded forest areas, such as the Tijuca National Park, an urban forest in Rio de Janeiro City. The Forest Nursery of Rio de Janeiro Botanic Garden grows around 490 tree species and makes 380 of them available for restoration and replanting schemes and for adding to the city's beauty.

The golden lion tamarin is an endangered species of the Atlantic Forest. Reserves established for charismatic animals such as this help protect the wide range of other species that share their habitat.

Daniela Zappi, a cactus specialist who grew up in Brazil and now works at Kew, considers that the best option for the conservation of the rainforest cacti is the protection and maintenance of their remaining natural habitats, but there is room for a combination of improved growing skills and eventual reintroduction to the wild of species whose populations are dwindling or have already disappeared from their native environment.

## RIO'S ORCHIDS

The first Brazilian botanist to be interested in native orchids was João Barbosa Rodrigues. Born in 1842, the son of a Portuguese trader and a Brazilian woman of Indian descent, he become director of the Rio Botanic Garden at the time it took on its scientific role in botany in 1890. He described 381 species and 11 new genera between 1877 to 1882. He travelled extensively throughout Brazil in his searches for orchids and illustrated almost 600 species. Today, the garden has over 800 orchid species in its collection. Magnificent species include the fragrant-flowered *Laelia lobata*, which grows only on two mountain tops, Pão de Açúcar (Sugarloaf Mountain) and Pedra da Gávea. Over-collection has been a major threat to the orchid at both these sites and it may well be extinct at Pão de Açúcar. Conservation efforts for *Laelia lobata*, one of 24 endangered orchid species of Rio de Janeiro City, include detailed surveys of the wild populations, micropropagation and education of climbers at the two known sites.

## SHARING EXPERTISE

Rio Botanic Garden has taken on the task of coordinating the development of a botanical checklist for the whole of Brazil. Another major challenge is the establishment of the National Center for Flora Conservation – CNCflora. This new centre will coordinate conservation action for species included in Brazil's official Red List, managing an integrated database with many other partners. These measures are a matter of urgency: the official list of endangered species of Brazilian flora, published in 2008, includes 472 threatened species and 1,079 species that are 'data deficient' (those for which there is not enough reliable information to make an assessment).

Throughout the country 34 botanic gardens are members of the Brazilian Botanic Gardens Network, which was set up in 1991 with the aim of creating centres of biodiversity conservation by sharing ideas and expertise. The main living collections held by the Brazilian gardens are regional, reflecting the local floras. Their value is increasing as agricultural practices and urban growth are degrading natural habitats and global climate change begins to take its toll. In addition most gardens, like Rio, also maintain nature reserves so there is great scope for integrated conservation. In 2004 the network produced an action plan that links to international targets for species conservation.

When the network produced its action plan it was agreed that a target of conserving 60 per cent of threatened native species in *ex situ* collections, in line with the Global Strategy for Plant Conservation, was too ambitious. The sub-target of protecting 90 per cent of critically endangered species was also felt to be too ambitious and so the goal of protecting 50 per cent of critically endangered species in living collections was adopted instead. Efforts to record and share the plants in Brazil's botanic garden collections are supported by a plan to computerize plant records into a common system, which will provide information helpful for recovery and restoration efforts for Brazil's most endangered plants.

Working together the botanic gardens of Brazil are a strong force for plant conservation. However, most of the gardens are located in the south-east of the country and there is a need to expand the network so that all the regions of Brazil have an equally strong botanical conservation focus. As the network grows in strength its vital role in protecting and restoring Brazil's rich flora will be realized.

China's vast flora is well documented as a result of the Flora of China project, an international collaboration between Missouri Botanical Garden, the Royal Botanic Gardens of Edinburgh and Kew and the Chinese botanical network. However, collecting detailed data on species distribution and conservation status still remains a challenge for this huge country. Rapid economic development in the last thirty years and continuous population growth seriously threaten the abundant plant diversity and nearly 4,000 to 5,000 higher plants are considered at risk of extinction, 15–20 per cent of all China's plants. In many cases species are known from just a handful of sites and further exploration, particularly in remote regions, is needed to find out how widespread these rare plants currently are.

**Previous page:** *Bretschneidera sinensis*, a forest species of China, Thailand and Vietnam. In China it is threatened by habitat destruction.

China has around 160 botanic gardens and arboreta designated by the government as research and development centres for plant conservation and sustainable development. Most are located in central city locations or in the suburbs, forming an integral part of the city environment. The number is increasing with major new gardens being developed or planned, for example the magnificent Chen Shan Botanical Garden in Shanghai, Nan Shan Botanical Garden in the ancient city of Chongqing in Sichuan Province and Dongguan Botanical Garden in Guangdong.

The current development of botanic gardens and the resurgence of interest in their potential within China is due mainly to the recent rapid development of the country's economy and the growth of external and internal tourism. Botanic gardens are seen as prestigious urban development projects and are encouraged by an ordinance of the Chinese government that emphasizes that all cities should build botanic gardens to conserve their local biodiversity.

## SOUTH CHINA BOTANIC GARDEN

Guangzhou, the sprawling city often referred to as Canton, is a port
connected to the South China Sea by the Pearl River. In the centre of the
metropolis is 'the forest of the city', the 330-hectare South China Botanic
Garden. This landscaped garden, the lungs of Guangzhou, is a major
recreational attraction but also fulfils the important functions of botanical
research and conservation.

The Director of South China Botanic Garden, Professor Huang
Hongwen, also chairs the Botanic Garden Committee of the Chinese
Academy of Sciences (CAS). He is passionately committed to the
conservation of Chinese plant diversity in a way that will benefit rural
communities within the country. 'During my lifetime I have seen massive
changes in the rural landscape of China. We urgently need to protect the
natural habitats that are left for the sake of our ecological and social well-
being. The natural plant wealth should also be fully explored for useful new
foods and medicine that can be developed to improve rural livelihoods.'
Professor Huang considers that the botanic garden network is essential for

The South China Botanical Garden
has many trees in its collection
including the largest *ex situ*
collection of threatened magnolias.

# CHINESE PLANTS IN WESTERN GARDENS

China's 31,000 species of vascular plants account for 10 per cent of the world total, making it one of the richest countries in terms of plant diversity. Many familiar garden plants have their origins in the temperate regions of China. Magnolias, rhododendrons, camellias, lilies and primulas are just a few of the plants introduced to gardens of Europe and America by plant explorers in the 18th and 19th centuries. Augustine Henry, an Irish doctor employed by the Chinese Imperial Maritime Customs Service, was one of the prolific plant collectors of the time. Working on behalf of Kew between 1881 and 1900 he sent

*Camellia chrysantha* is threatened by habitat loss in its native China and Vietnam. Fortunately it can be propagated readily from seed or cuttings and by tissue culture.

158,000 dried specimens to the UK, including 500 new species, together with bulbs and seeds of many Chinese plants. He was deeply concerned about the loss of forests and their plant species near the Chinese border with Vietnam – an area now recognized as a global biodiversity hotspot. Even in Henry's day timber felling and charcoal production were major threats to the flora. Henry collected seed of the handkerchief tree, *Davidia involucrata* and sent this to Kew, but Ernest H. Wilson was the collector who ensured this plant was established in cultivation in the west. Wilson also introduced various species of honeysuckle, rose, maple and clematis, *Actinidia chinensis* (Wilson's gooseberry) and the spectacular *Lilium regale*. Wilson considered that he 'would proudly rest his reputation with the Regal Lily' that he found growing on mountain slopes in the province of Sichuan. *Lilium regale* has a limited natural distribution but with its fragrant white flowers has become hugely popular in cultivation.

Nowadays many of the plants collected in the era of Henry and Wilson are threatened with extinction in the wild. The plants growing in botanic gardens both in China and elsewhere in the world offer the potential for restoration of wild populations.

research and for storing China's plant diversity, which is being so rapidly eroded in the wild. 'The botanic gardens are where we can study plant diversity under controlled conditions and maintain long term *ex situ* collections either as living plants or in seed banks. As elsewhere in the world we need to be vigilant in documenting our conservation collections and carefully recording the origin of wild source-material so that we retain future options for reintroduction and ecological restoration. More importantly, the botanic gardens can serve as regional and national bioresource reserves and exploration platforms for the newly emerging bioeconomy.'

The 16 botanic gardens of the Chinese Academy of Sciences have about 20,000 vascular plant species in their collections. The three core gardens, South China Botanic Garden, Xishuangbanna Tropical Botanical Garden in Yunnan and Wuhan Botanic Garden, account for approximately 90 per cent of all plant species maintained by Chinese botanical gardens and through these the Global Strategy for Plant Conservation target of conserving at least 60 per cent of native plants has been achieved. The living plant collections are complemented by a major new seed bank, the Germplasm Bank of Wild Species opened at the Kunming Institute of Botany in 2008. The goal of this new seed bank is to collect and conserve 6,450 species within five years and within fifteen years to have seeds of 19,000 species in storage.

## MAGNOLIAS

South China Botanic Garden has an extraordinary diversity of plants within its collections. Designed mainly for *ex situ* conservation, over 13,000 plant taxa are grown in over 30 specialized gardens, including areas for magnolias, palms, gingers and orchids along with the medicinal plant garden. For over 50 years there has been an emphasis on the study and use of the tropical and subtropical plant resources of South China, and under the leadership of Professor Huang, conservation of rare and threatened species is becoming increasingly important. The range of magnolia species in cultivation is particularly impressive. The garden has the richest collection of magnolias in the world with 86 magnolia taxa represented from various different countries. From a conservation point of view the most valuable holdings are the specimens of 10 Critically Endangered and 13 Endangered Chinese species of magnolia. These need to be very carefully managed as a precious conservation resource.

Professor Huang Hongwen, Director of South China Botanic Garden, plays a major role promoting the conservation of China's threatened plants.

Southern China is the world centre of diversity and distribution of magnolias with over 40 per cent of the 242 known species occurring there. A significant number are considered globally threatened because of habitat decline and in some cases over-exploitation. Certain species are collected for the medicinal products they yield and others are harvested from the wild for planting in gardens and as street trees. Habitat fragmentation is the most serious problem leading to increasingly isolated remnant populations of magnolias that often struggle to regenerate under natural conditions. Some species are also affected by the loss of their wild pollinators.

Botanists from South China Botanic Garden are currently working to collect information on a range of priority species of magnolias in the wild – studying and recording their ecology, conservation status and the threats they face. Armed with this information they are strengthening *ex situ* conservation efforts and looking at the possibilities for restoring natural populations. The work is part of a global effort to save tree species from extinction through the Global Trees Campaign – a collaborative programme between BGCI and the wildlife conservation organization, Fauna and Flora International.

**Left to right:** *Magnolia zenii, M. ventii, M. longipedunculata.*

The critically endangered *Magnolia hebecarpa* is currently only known to be in cultivation at the South China Botanic Garden. In 2008 botanists found a single mature specimen of the species during field surveys in Hekou County, Yunnan. Approximately 300 seeds were collected from the tree to supplement *ex situ* collections. In time it may be possible to work with local people to replant young trees in Hekou County and help ensure the long-term survival of the species. Fortunately a new population was also discovered in the Daweishan Nature Reserve, which is located in the southeast frontier of Yunnan close to the border with Vietnam. A very important step in the conservation of *Magnolia hebecarpa* is to make people aware of the global rarity of the species, which is unique to the province of Yunnan.

Two other Endangered magnolias endemic to Yunnan were found to be in only one *ex situ* collection in China. South China Botanic Garden is taking steps to boost conservation efforts for both these species: *Magnolia ingrata* and *Magnolia ventii*. During field work in 2008, 12 mature specimens of *M. ingrata* were discovered in Maguan County, Yunnan, in highly disturbed forest habitat. Seed was collected so that additional plants can be produced in

cultivation. Again, the longer-term objective is to restore wild populations of this highly threatened species.

A fourth magnolia that South China is taking steps to conserve is the recently discovered *Magnolia longipedunculata*, which was only described in 2004. Field surveys in three counties of Guangdong have identified only a single population of 11 mature *M. longipedunculata* specimens and the species is considered to be critically endangered. Scientists are studying the ecology, phenology and reproductive characteristics of the wild population of this species so that they can work out the best way to conserve it.

In general, magnolia experts around the world have found their efforts to propagate the rarer species of magnolia limited by poor germination and establishment rates. More research is urgently required to understand seed production and propagation in all threatened taxa in order to support of the establishment of *ex situ* collections as well as restoration and reintroduction activities.

Seed heads of *Magnolia longipedunculata*. An understanding of seed production and germination is an important part of work to conserve this Critically Endangered species.

## OTHER CONSERVATION PROJECTS

South China Botanic Garden's plant conservation work extends beyond the magnolias. Integrated approaches link research with conservation in natural habitats as well as propagation and ex situ conservation within the garden. This approach is being taken, for example, with *Bretschneidera sinensis*, a tree species that is considered to be endangered in southern provinces of China. A relict species of the Tertiary era, it is under threat because of habitat destruction, low seed production and problems in natural regeneration. The stunning flowers of *Bretschneidera* make is a potential candidate for growing as a garden ornamental, potentially adding another Chinese species to the global garden flora.

With seed collected from six wild populations, germination trials have been carried out and *ex situ* populations established in South China Botanic Garden. Plants are also being grown at a rural propagation centre. An important locality for this magnificent tree is Nankunshan Provincial Nature Reserve. Field pollination trials have been established in Nankunshan, to help work out how

# MEDICINAL PLANTS

As well as a wide range of ornamentals, China has a huge diversity of medicinal plants. These are of primary importance for healthcare throughout the country. Over 11,000 species are used in traditional Chinese medicine. Of the 600 plant species that are regularly used, about one third are well-established in cultivation, which helps to prevent unsustainable harvesting from the wild. Nevertheless, wild harvesting of medicinal plants for local consumption and international trade continues to be a major threat to wild plant diversity in China. Botanic gardens have an important role to play in increasing the range of species in cultivation. South China Botanic Garden has a medicinal plant collection that includes about 1,000 species.

Looking ahead, medicinal plants are expected to form part of China's bioeconomy. Sustainable use policies will be a vital part of strategies to benefit from this rich natural resource.

Left: *Tigridiopalma magnifica* growing on the forest floor. This species is Critically Endangered but measures are now being taken to boost the size of the population in the wild using plants grown by micropropagation techniques.

Right: The Critically Endangered *Primulina tabacum* in its natural habitat. Scientists have studied the genetics of the species to aid its long-term conservation.

seed production can be improved. Working with the local communities is very important for the long-term conservation of *Bretschneidera*. A workshop was held in September 2008 within the Nature Reserve to explain the importance of the species and engage local people in efforts to conserve it. The response was very encouraging with people keen to help when they understood the global rarity of the species. Fortunately this forest tree species might be less threatened in the wild than previously thought. It has been found recently in Taiwan and Thailand.

Another species being targeted for conservation attention is the critically endangered *Tigridiopalma magnifica*. With attractive foliage and beautiful flowers, this plant has considerable potential for as an ornamental pot plant. Scientists studying taxonomy and plant diversity are also particulary interested in the species because it has no close relatives; it is the only species in its genus. Scientists have collected plants of *T. magnifica* from E'fengzhang Nature Reserve, Yangchun County, and other localities. Seed germination trials have been carried out and micropropagation techniques developed to bulk up the species. To date, more than 1,000 plants have been reintroduced to the wild.

*Primulina tabacum*, a Critically Endangered perennial only found in limestone areas in northern Guangdong is also the focus of research at South China Botanical Garden. The garden's scientists have studied the genetic variability within and among four populations of this species and have discovered that there is little gene flow between populations. To maintain present levels of genetic diversity and increase the potential for long-term survival – which may require adaptation to a changing climate – *in situ* conservation of all populations is necessary. This species is already grown by enthusiasts and has considerable horticultural potential.

As well as single species conservation the South China Botanic Garden is responsible for the management of its own protected area and three field stations. The Dinghushan National Nature Reserve, situated in the northeastern suburb of Zhaoqing city, 84 kilometres away from Guangzhou, was established in 1956 as China's first national nature reserve. In contrast to the surrounding areas of the densely populated Guangdong Province, the Dinghushan Reserve comprises relatively pristine primary forest. It is internationally recognized as a UNESCO Biosphere Reserve and is also a place of pilgramge with many Buddhist temples. The reserve has up to a million visitors each year. Within an area of 1,133 hectares the uninhabited reserve gives *in situ* protection to over 2,400 plant species in a range of vegetation types including monsoon evergreen broadleaved forest. The dominant tree species include the Kweilin chestnut (*Castanopsis chinensis*), *Schima superba* and *Engelhardtia roxburghiana*. Almost half of the plants in the reserve have been recorded as having some medicinal use. Scientists at a research station within the reserve are working to strengthen understanding of ecological processes within the forest and the potential for ecological rehabilitation and reconstruction. They are also actively involved in assessing the best methods for the management and sustainable utilization of forest resources – essential in China with its rapidly growing population. A major study on carbon cycling by subtropical forest ecosystems is also taking place in Dinghushan, which will assess the carbon contribution coming from the southern subtropical ecosystems to the total carbon sink of China.

## PLANNING FOR THE FUTURE

South China Botanic Garden's work in plant conservation will increase in the future in response to national and global challenges. Professor Huang was

**Above:** China has a rich diversity of native rhododendrons with over 550 species. In some forest habitats, rhododendrons are the dominant trees and shrubs, as in this forest in Guizhou.

**Right:** Rhododendron seedlings in Lushan Botanic Garden. Propagation of the rarer species in botanic gardens is an important component of the long-term conservation of rhododendrons. Lushan Botanic Garden grows a wide range of species.

one of the architects of China's Strategy for Plant Conservation, the national response to the Global Strategy for Plant Conservation. He sees this as a very important blueprint for conserving China's huge diversity of native plants: 'As China grows in prosperity our commitment to biodiversity conservation will develop further. Botanic gardens are key to the successful conservation of plants in China. We have *ex situ* plant resources and scientific expertise that is committed to developing conservation solutions. We now need to reinforce coordination and cooperation amongst botanic gardens and develop closer links with *in situ* conservation agencies. China's Strategy for Plant Conservation unites the different organizations in a common plan to save 10 per cent of the world's flora.'

One of the most celebrated of Asia's gardens is the Bogor Botanic Garden on the Indonesian island of Java. Close to the densely populated capital of Jakarta, the Bogor Botanic Garden is situated on the slope of Mount Salak, an eroded volcanic range. Bogor has cooler weather than Jakarta and for this reason the Dutch Governor General Baron Gustaaf von Imhoff chose to build a summer house there in 1744. For two centuries the Dutch rulers remained in residence at this site, replacing the summer house with a more imposing palace. For a brief period the British occupied Java and during their stay a botanic garden was created and officially opened to the public in 1817. The Dutch nurtured the garden on their return and it became an important research centre for tropical crops such as cassava, tea, oil palm and cinchona. The garden also developed as a fine landscaped park for the enjoyment of the colonial elite.

**Previous page:** Lush vegetation at Cibodas Botanic Garden.

The Bogor Botanic Garden provides the headquarters for Kebun Raya Indonesia, a complex of four major gardens stretching from West Java to Bali. Collectively these gardens house the most extensive living collection of plants native to Southeast Asia and are the botanical ark for an area of huge diversity. Approximately 29,000 native plants have been recorded from Indonesia alone and many of these species are under threat in the wild as rainforests are progressively degraded and destroyed. On the island of Java, one of the most densely populated parts of the world, virtually all the lowland rainforest has long been cleared and replaced by intensive cultivation.

Indonesia has been recognized as one of the world's 25 biodiversity hotspots. At least 590 plant species of this region are classified as threatened by the IUCN. The real figure is likely to be considerably higher and will increase as more species are assessed. The plant collections in the four major Indonesian gardens represent approximately 10 per cent of the native species. The ability to electronically database and determine which collections represent rare, endangered or endemic species was developed about ten years ago in Kebun Raya so that the network can now effectively monitor its conservation work.

## CIBODAS, PURWODADI AND EKA KARYA BALI

Two of the other gardens in the Kebun Raya Network were created as experimental agricultural and horticultural gardens by the Dutch. In the mountains above Bogor, an area of cooler climate and heavy rainfall, is the botanic garden of Cibodas. Established in 1862, this is considered by many to be one of the most beautiful botanic gardens in the world. Cibodas is situated at an altitude of nearly 1,500 metres on the slopes of mount Ged-Pangrango and is able to grow many temperate species. The garden covers an area of 125 hectares, about one third of which remains forested.

Purwodadi Botanic Garden was established in 1939 in Lawang, East Java, an area which formerly had monsoon forest vegetation. Now Purwodadi Botanic Garden has an impressive collection of around 400 species of orchid. One of these is the endemic and globally endangered *Paphiopedilum glaucophyllum*, a slipper orchid that is used as the garden's symbol. Eka Karya Bali Botanic Gardens is the youngest of the four gardens, and the only one established by Indonesians following independence – it was founded in 1959 on the site of a former timber concession.

Natural vegetation in Java. Forests are now mainly confined to mountain areas, where they continue to provide timber, fuelwood and medicinal plants for local people.

# THE RECORDS CHALLENGE: DUSTY RECORDS TO MODERN DATABASES

The development of the Kebun Raya collections database was a massive undertaking carried out over three years from 1995 to 1998, as a jointly managed project between Kebun Raya, BGCI, and the Indonesian Network for Plant Conservation (INetPC). The consolidated database for all four gardens is maintained in Bogor.

The project had to contend with a range of historical problems. Early plant collection data from the 19th century were scattered through dusty old record books written in Dutch in fading ink and containing minimal information about species names and locations. Many plants in the gardens were found to be mislabelled or to have no visible tags. Visiting botanists worked with Kebun Raya staff to check and revise outdated or inaccurate plant names. In all four gardens, the orchid houses, seed banks, herbaria and speciality greenhouses such as the fern and succulent houses at Cibodas Botanic Garden had separate record systems. Bringing all this data together was complex to say the least. The painstaking process and at times frustrating progress working with nearly two hundred years' worth of accumulated botanical knowledge were ultimately worthwhile. Computerizing the Kebun Raya collections laid the basis for planning *ex situ* plant conservation on a national scale for Indonesia – one of the most biodiverse countries on earth and a flora that the world cannot afford to lose.

The flower of *Rafflesia*, a parasitic plant, smells of rotting flesh. All species occur in the Indonesian rainforests, and their future is dependent on habitat conservation.

Nowadays, Bogor Botanic Garden and its sister gardens have very important roles to play in plant conservation, education and taxonomic research. This is vitally important work in a country which is so rich in plant diversity and which is experiencing rapid rates of forest loss. The Bogor Botanic Garden itself has over 15,000 plant species within its grounds, including plants from throughout Southeast Asia. Part of the garden has a collection of huge, old dipterocarp trees – the mainstay of Indonesia's timber industry. Over 30 of the dipterocarp species grown at Bogor are considered to be globally threatened by IUCN. These include *Hopea bancana* and *Hopea nigra*, both Critically Endangered species that are endemic to Sumatra.

Orchids, timber trees and palms are amongst Indonesia's most endangered plant species. Over 90 orchids and 31 palm species are considered to be threatened with extinction. In each case the true number of threatened species is likely to be much higher. Indonesia has over 470 species of palms, more than any other country in the world. Around 10 per cent are used either for subsistence or for commercial purposes. Some are widely cultivated and are

available commercially, including the ornamental sealing wax or lipstick palm (*Cyrtostachys renda*) that is native to Malaysia and Sumatra, sugar palm (*Arenga pinnata*), and the staple food of Irian Jaya: sago (*Metroxylon sagu*). Wild rainforest palms are used throughout the country, and the climbing palms or rattans are very important for the production of furniture. Various species of rattan are under threat in the wild as a result of collection for trade. One species, *Calamus spectabilis*, is very possibly extinct as there have been no records in the last few decades and another valuable rattan, *Calamus manan*, is now Endangered in the wild. Bogor Botanic Garden is helping to conserve this species through propagation and reintroduction to the Bukit Dua Belas National Park in Sumatra.

## PINANGA JAVANA

Among the ornamental palms around 40 species of *Pinanga* grow in Indonesia, including 14 which are endemic to the country. Many of these species are under threat in the wild. One species, known only from one hillside location, has been cultivated by Bali Botanic Garden for many years. It was recently described for the first time as *Pinanga arinasae*, named after a former head of the garden.

*Pinanga javana* is another very rare and endangered species, endemic to Java. As well as being an attractive ornamental the stem of this species is used locally for constructing the traditional open dwelling or 'pondok' and the young stem apex is sometimes eaten as a vegetable. The decline of this species is closely associated with human activities, particularly habitat destruction and conversion. Large areas of its former habitat have been considerably altered by human settlement, crop plantations and the development of large timber plantations, especially of *Pinus merkusii* and *Agathis borneensis*. Collection for horticulture and browsing deer and monkeys are additional minor threats.

*Pinanga javana* is cultivated by Bogor Botanic Garden, Cibodas Botanical Garden and the newly established Baturrarden Botanic Garden. This new garden is located on the slopes of Mount Slamet in Central Java. Although not officially part of the Kebun Raya system (it is governed by the local provincial government), it receives advice and technical resources from Bogor Botanic Garden. Dr Didik Widyatmoko, Director of Cibodas Botanic Garden, is leading attempts to conserve *Pinanga javana* and reintroduce it back into native forest habitats with the hope that the palm can become established in the wild and maintain its genetic diversity. Over 5,000 individuals have been reintroduced into 16 locations in Gunung Halimun National Park and surrounding areas in West Java.

Dr Widyatmoko recognized that enhanced *ex situ* protection and recovery planning was urgently needed for this important palm. He considers that this species can be used as a flagship species to promote conservation of the Javanese lower montane ecosystems but he recognizes the challenges of studying and successfully restoring the palm. '*Pinanga javana* is a solitary and slow growing species and appears to require precise environmental conditions for regeneration. Flowering and seed production vary markedly from year to year. Heavy flowering does not always mean good seed production as prolonged wet weather during or immediately following flowering can lead to the flowers rotting and falling off the plant. Pollination systems of *Pinanga javana* are not well understood but based on my personal observations, the species seems to be predominantly insect-pollinated. Beetles and weevils seem to be the important pollination agents, while birds may play a minor role.'

Mount Salak, rising above Bogor, has just two remaining populations of *P. javana*, including the largest known wild population, growing on steep rocky slopes in a gorge. When John Dransfield, a leading palm expert based at Kew, visited the site in 1971 around 1,000 plants were present, but following a serious decline fewer than 300 plants now remain. Active management is needed to adequately protect this crucial population. The second site at Mount Salak has only two adult individuals on a very steep hillside vulnerable to landslides. A small number of adult plants occur in montane forest above Cibodas Botanic Gardens. The first important step in managing the reintroduction is to conduct detailed checks establish where *P. javana* is still found wild. To be fully effective, the recovery plan, and indeed restoration plans for Indonesia's many other threatened species, has to be integrated with broader conservation strategies. Encouraging public involvement in plant conservation is extremely important and here the country's botanic gardens have an important role to play.

**the bryophyte garden at Cibodas helps to focus attention on these important plants**

## CONSERVATION PLANNING

Spectacular ferns are a particular feature of the cool lush gardens, and Cibodas is successfully propagating threatened tree ferns for restoration into the wild. Indonesia is particularly rich in moss species with over 3,000 different species, 250 of which are found in the Cibodas Botanical Garden. Mosses receive very limited conservation attention and it is hoped that the bryophyte garden at

The bryophyte garden at Cibodas is a small area devoted to the cultivation and display of different types of moss and liverwort.

Cibodas will hlep raise awareness of these ecologically important plants as well as increasing knowledge of their cultivation.

Looking ahead, the Kebun Raya is set to grow and more botanic gardens are planned to help study and conserve Indonesia's rich flora. Collaboration with overseas gardens is also important. In one major collaborative venture a few years ago the Royal Botanic Garden Edinburgh repatriated some fourteen Indonesian rhododendron species to the Cibodas Botanic Gardens, including all their associated collection data. These tropical 'vireya' rhododendron species were collected from Indonesia as live material in the past, and had never been cultivated in Indonesia. In order to ensure that the botanic gardens staff learned specific techniques for rhododendron collection, propagation, and horticulture, a technical workshop was held at Cibodas, organized by Kebun Raya and BGCI.

Dr Peter Wyse Jackson worked for many years with Indonesian botanic gardens during his leadership of BGCI. 'The traditional collaboration and cooperation between botanic gardens throughout the world has strengthened and expanded very significantly over the last decades,' he says. 'This is particularly helpful in ensuring that the much greater botanic gardens resources

Tree ferns are being cultivated in Cibodas Botanic Garden for reintroduction to the wild. Local people dig up *Cyathea* from the wild to sell to visitors to the National Park close to Cibodas. Working with local communities, propagation will provide a more sustainable source of these plants.

available in many temperate countries can be used to support the efforts of our colleagues in many tropical countries. The ways in which tropical and temperate botanic gardens can collaborate are many and varied. Sharing expertise in areas such as plant taxonomy, horticultural management, environmental education techniques and project management are just a few of the ways where temperate and tropical gardens can work together. Temperate gardens also have an important role to play  in providing staff training places for their colleagues from tropical gardens and in supporting their fund-raising efforts too.'

The resurgence of interest in botanic gardens in Asia, fostered by China, is likely to ensure that in the future tropical botanic gardens will increase their role in plant exploration and conservation.

The small, landlocked country of Uganda has an extraordinary diversity of plant and animal life. The richest country for birds in Africa, Uganda has almost half the species known on the African continent and over 10 per cent of the world's avifauna. Over 1,000 species of birds have been recorded in the country. The flora consists of around 5,000 species with most of the plants shared with neighbouring countries and relatively few endemics. Most of the country is covered in a mosaic of secondary grassland and farmland with patches of remnant forest and swamp. In the west of the country is tropical rainforest – the eastern extension of the dense equatorial forests of the Congo Basin.

**Previous page:** Lake Bunyonyi, in south-west Uganda. Terraced farmland has replaced much of the natural vegetation but forest patches are important for local livelihoods.

Much of Uganda's vegetation would revert to some form of forest in the absence of fire and cultivation. Agriculture was introduced around 2,500 years ago and since then the process of clearance for crops and pasture has continued apace. Now 80 per cent of Ugandans are farmers growing a mix of crops for subsistence, with bananas being the staple crop. There is still a huge reliance on wild plants extracted from the remnant forests, grasslands and wetlands and the demand for fuelwood is enormous. This is partly supplied from plantations but also from the natural forests and woodlands. Charcoal is sold all along the few major roads in the country and the demand for building poles is also substantial. Raffia (Raphia farinifera) and bamboo (Arundinaria alpina), are collected to make baskets and mats – among the most significant handicrafts produced in Uganda. Wild plants are also harvested on a major scale for traditional medicines, which provide the only form of healthcare for the majority of people in Uganda.

Fortunately most of the rich forests of Uganda are now under some form of protection. The development of forest reserves began over one hundred years ago when the colonial government signed an agreement with the rulers of

The office of the Nature Palace Botanic Garden. The garden works very closely with local communities to supply medicinal plants.

Uganda's ancient kingdoms to bring the forests under national control. The forest reserves were created primarily to satisfy the country's wood requirements. In addition to forest reserves, other areas of forest are protected in national parks and nature reserves. Although forest now covers only about four per cent of the country, the general rate of deforestation is relatively low. But agricultural encroachment continues to be a problem, for example when immigration occurs from war-torn neighbouring countries. Innovative solutions are needed to improve the livelihoods of rural people without destroying the biodiversity on which they depend. In the west of Uganda, nestling in the foothills of the Ruwenzori Mountains, Africa's tallest mountain range, a new botanic garden is helping to demonstrate how this may be possible.

## TOORO BOTANICAL GARDENS

Tooro Botanical Gardens, situated in the town of Fort Portal, was established in 2001 on a site formerly managed as the Njara Forest Reserve by the Ugandan National Forest Authority. Fort Portal is the capital of the traditional kingdom of the Tooro people. The ancient kingdom was reinstated in 1993 and is now ruled by King Oyo Nyimba Kabamba Iguru Rukidi IV who lives in an imposing palace in the town. Professor Edward Rugumayo, former Minister of Internal Affairs, lives close to the garden and realized the huge potential of the Forest Reserve site as a conservation centre for the plants of the rich local rainforests and beyond. Kabaseke Clovis, a forester and environmental campaigner, agreed to support the idea of creating a botanic garden and in 2004 the garden was officially registered as an NGO. One of the first tasks undertaken was to create small nurseries to show that the land was occupied. Long-term work also began to remove many of the eucalyptus trees that had been planted at the Njara Forest Reserve since the early 1950s and gradually to replant the area with indigenous tree species. One historical stand of eucalyptus planted in honour of a visit by Princess Elizabeth prior to her coronation will, however, be retained as a source of seed for fuelwood plantations.

Plants for sale at Tooro Botanic Garden. The garden is demonstrating methods of vegetable production to improve local nutrition and works with local people to secure the conservation of medicinal plants.

The overall impression of Tooro Botanical Garden is one of lush greenery. The fertile soils and abundant rainfall allow a rich mix of plants to grow. Now after eight years the garden has patches of cultivated ground and areas of tangled semi-natural vegetation. Trails through the garden with lookout posts at treetop height have been created for the increasing numbers of visitors to this wonderful tropical garden.

Visitors can learn a lot from their time here. Tooro Botanical Garden is committed to demonstrating ways of conserving biodiversity and at the same time improving the supply of important plants – and income – for local needs. The medicinal plants particularly valued by local people have been identified and documented through community meetings. Now the garden is working with around 70 local women on a voluntary basis to grow local herbs and spices and has produced a book to show how these can be used medicinally to improve family health. A model farm has been developed within the botanic garden showing how medicinal plants grown for family use can be planted between cash crops. Every two months traditional healers come together at the garden to discuss specific health topics.

The garden has also begun drying and packaging medicinal herbs for sale. So far these include non-native species such as wormwood (*Artemisia annua*), a plant with great promise in the battle against malaria, *Justicia adathoda* used to treat measles, flu and respiratory problems and gotu kola (*Centella asiatica*), a memory booster. Income from the sale of these products helps to pay the running costs of the garden and to provide employment for people in the Fort Portal area.

## KIBALE FOREST NATIONAL PARK

Western Uganda has a rich diversity of tree species now mainly confined to the forest reserves and other protected areas. Kibale Forest National Park, close to the Tooro Botanical Garden, has around 350 tree species. Various types of forest habitat occur in the park with evergreen rainforest along the Fort Portal plateau and woodland and savannah along the Rift Valley floor. In the central part of the park, around Kanyanchu, the high forest consists of a mixture of evergreen and deciduous trees with a canopy 55 metres high. The undergrowth has a variety of shrubs, ferns, and broad-leaved forest grasses, many of which are valued by the local people for a range of daily uses.

Kibale Forest, one of the most important biodiversity sites within the Albertine Rift, is also home to a remarkable 13 primate species, including chimpanzees, the endangered Ugandan red colobus monkey and vulnerable L'Hoest's monkey. The forest elephant, smaller and more hairy than its savannah relatives, moves seasonally into the national park, and other mammals include buffalo, giant forest hog and antelope species. The forest is home to 335 bird species including the endemic Prigogine's ground thrush.

Tooro Botanic Garden has ambitious plans to conserve the trees of the Kibale Forest and extend their cultivation and use for the benefit of local people. In the future it may be possible to utilize a much more extensive range of trees than the ubiquitous pines and eucalypts which are still routinely planted in Uganda. Quentin Meunier, a French botanist working at Tooro Botanic Garden, has been involved in promoting the cultivation of a wider range of trees within the region. 'Our surveys have shown that there is a really strong interest amongst farmers and traditional healers living near Fort Portal to grow native trees in their home gardens. It is so important that we

*Justicia dathoda* (left) and *Artemisia annua* (right) are two medicinal plant species grown, dried and packaged for sale at Tooro Botanic Garden.

# THE ALBERTINE RIFT

The Albertine Rift Mountains covering south-west Uganda, Rwanda, Burundi, and a tiny part of Tanzania and eastern Zaire form a global biodiversity hotspot. Various kinds of forest clothe the mountains and Afro-alpine moorlands with giant lobelia and senecios are found at higher altitudes. Glaciers still remain but these are retreating as a result of global climate change. Around 800 species of plant have been recorded in the Albertine Rift, accounting for about 14 per cent of the entire African flora. Rich in birdlife, 1,061 species of bird have been recorded in the region, 25 of which are considered to be highly threatened, and 402 species of mammal have been recorded. Thirty-five mammal species are considered to be highly threatened, including the mountain gorilla, eastern lowland gorilla and chimpanzee.

Conversion of forest areas in the Albertine Rift for farm land, tea plantations, and plantations of exotic trees is depleting plant resources and threatens the integrity and survival of the remaining ecosystems. Loss of vegetation also threatens water supplies, with the snow-capped Ruwenzori Mountains, the fabled Mountains of the Moon described by Ptolemy, forming one of the most significant catchment areas in Uganda, and the most permanent source of the White Nile.

show how this can be a reality. Nearly all trees have medicinal values for local people as well as a variety of other uses. We cannot afford to lose this knowledge or the tree species themselves.'

## TREE CONSERVATION AT TOORO

Tooro's work involves locating and documenting the trees within Kibale Forest and collecting material for cultivation in the garden. So far over 60 trees have been established in an arboretum within the garden. One very important tree in cultivation is *Warburgia ugandensis* commonly known as East African greenwood or in local languages as medicine tree. The species has become very scarce in the wild with only scattered individuals found in Kibale and other local forests. The tree is prized by traditional healers for its antibiotic properties, and is used as a painkiller to treat a variety of diseases. The tree also produces useful timber for furniture and edible fruits. Seeds from *Warburgia* – when they are available from the scarce wild trees – cannot be stored and must be sown straight away. Propagation using stem cuttings is likely to be a more effective means of producing trees in cultivation.

Another tree being grown at Tooro is *Beilschmiedia ugandensis* an uncommon species in western Uganda that is restricted to areas deep within the reserves and inaccessible to rural communities. Recent domestication trials have shown that this species could be important in farmed landscapes providing fuelwood, charcoal and edible fruit.

On a larger scale, cultivation of *Prunus africana* is very promising. Tooro has a small plantation of this evergreen tree, which grows up to 30 metres tall and has small white flowers bearing cherry-like fruits. After seven years the trees at Tooro are already beginning to look impressive. The rough thick bark of *Prunus africana* is dark brown to almost black. This is the part of the plant which is most prized. Throughout much of its range, and most notably in Cameroon and Madagascar, collection of the bark for the commercial production of a drug used to treat prostate cancer has led to serious declines. The tree has also been exploited for its valuable timber. Now the species is too rare in many countries to be of commercial use and is most commonly exploited for the local market. Tooro is demonstrating how the tree can be grown to supply its valuable medicinal products. *Centella asiatica* is being cultivated under the shade of the trees, providing an interesting mix of medicinal plant production in a relatively small space.

*Encephalartos whitelockii* is a species being taken care for by Tooro Botanic Garden not because of its current utility but because it is one of Uganda's relatively few endemic plants and a global rarity. This cycad, first described in 1995, is confined to the Mpanga Falls near Lake George to the south of Fort Portal. The entire population consists of around 10,000 plants that form a remarkable cycad forest. Professor Rugumayo and his colleagues at Tooro have been leading calls for the full protection of the site where *Encephalartos whitelockii* grows in the wild. This is urgently needed as work began on construction of a hydro-electric power facility at Mpanga Falls in 2008. Once the dam is finished the local microclimate may be significantly altered. As well as protecting the natural population, Tooro is successfully maintaining *ex situ* conservation collections of the cycad and has distributed plants to other botanic gardens in Uganda.

The very rare cycad, *Encephalartos whitelockii* in cultivation at Tooro Botanic Garden. This species is known from one small area in the wild. The site is being developed for hydro-electric power but so far, influenced by the botanic garden, the developers are helping to protect the cycad.

Botanists at Tooro are keen to pool their experiences and expertise with other botanic gardens in Uganda. The garden's director, Dr Godfrey Ruyonga, considers that, 'We have a huge responsibility to conserve the trees of the Ruwenzori foothills for the benefit of local people but we also need to help conserve medicinal trees of national interest. The best way to do this is to collaborate with Uganda's other botanic gardens, universities and other conservation agencies. Working together there is real hope for the future.'

## NATURE PALACE BOTANICAL GARDEN

Another young botanic garden in Uganda is the Nature Palace Botanical Garden in the Wakiso District close to the capital city of Kampala. Like Tooro, Nature Palace Botanical Garden is also being developed on a former forest reserve. The surrounding land is mainly farmland but there are some good remnant patches of forest close by. In 2005, Nature Palace Foundation started implementing a project which aims to strengthen community collaboration by cultivating medicinal plants within home gardens.  After consulting local people to find out which medicinal plants were becoming scarce a nursery was established so that seedlings could be distributed to the villagers to cultivate in their own plots. Nature Palace's community medicinal garden serves as a conservation store for declining species and a local centre for horticultural training. Local knowledge of medicinal plant use is supplied by Joseph Kgongo who is well-respected in the area for his expertise in identifying the plants and their uses.

Over 50 species are now grown in the home gardens – wild or naturalized plants that were previously harvested from the remnant natural forests and farmlands. Local people have formed the Twekembe Herbal Medicine Promoters Association and are involved in the processing of the medicinal products they harvest from their gardens. They have opened up a small distribution centre where they sell their herbal products mainly to local people.

Both Tooro Botanic Garden and the Nature Palace Botanic Garden are providing new models to show how important botanic gardens are in Africa for linking the conservation of plants with the needs of local people. They are involving local communities in selecting species for conservation action and showing how key species can be more widely grown in home gardens. Historically,  botanic gardens have been places for experimenting with the cultivation of plants for agricultural production, usually on a national scale. Now communities are becoming more directly involved.

The bark of *Prunus africana* is used medicinally to treat a variety of ailments and exported for use in making a drug to treat prostate cancer. Over-harvesting, as shown here, has led to serious decline across much of the tree's range.

## ENTEBBE BOTANICAL GARDEN

Agricultural development was the main role for Entebbe Botanical Garden when it was established over one hundred years ago. Situated on the banks of Lake Victoria, it was colonial East Africa's main garden for the introduction and evaluation of cash crops including cocoa, coffee, tea and rubber. Timber species were also evaluated and fine specimens of timber trees planted in the 19th century now provide shade in the impressive equatorial garden. At one time the garden had a collection of 2,500 species of plants of tropical, sub-tropical and warm temperature zones but the numbers dwindled due to neglect during the period of political turmoil under Idi Amin.

Now the collections are being upgraded and the scientific value of the garden reinstated. The Director, John Wasswa Mulumba, considers important objectives of the garden to include the screening of native wild plants for useful products and possible economic production in collaboration with other relevant organizations. 'Uganda's natural plant wealth is one of our major assets. We need to conserve and utilize our native plant resources for the benefit of Uganda's rapidly growing population. We have wild plants of great medicinal value, wild fruits, fibres, spices, gums, fuelwood and forage crops all worthy of scientific research and development. With our tropical climate, plants grow so well here and we can grow solutions to help overcome rural poverty.'

Entebbe Botanical Garden is carrying out research on a wide range of plants with potential value in the rural economy. *Artemisia* production is being studied with comparisons of the indigenous *Artemisia afra* alongside the naturalized *Artemisia annua*, both of which have an expanding role in the treatment of malaria. Indigenous fruit trees are also being studied. *Garcinia buchananii* is one such plant. Populations of this tree, valued both for its edible fruits and medicinal properties are decreasing mainly due to unsustainable harvesting methods and encroachment of forest reserves.

## LOOKING AHEAD

The wild plants of Uganda offer hope for the future. In a country self-sufficient in food and with an economy dependent on agriculture, wild plants offer scope for diversification and the development of new products. The botanic gardens both well-established and new are important centres for studying and looking after Uganda's native plant diversity.

A traditional healer presents plants in Bushenyi district (above) and checks a medicinal solution (below). Medicinal plants are of huge importance in Uganda where they are used alongside Western medicine. Some introduced species are gathered from former farmland where they grow as weeds but many species harvested from natural populations are declining in the wild.

Madagascar is universally considered to be one of the top priorities for biodiversity conservation. Since the island separated from the ancient landmass of Gondwanaland approximately 120 million years ago the plants and animals have evolved in relative isolation, resulting in many species that are not found anywhere else in the world. More than 90 per cent of Madagascar's plant species are unique to the island and for woody plants the degree of endemism is even higher, at 96 per cent. With dramatic escarpments, mountain ranges and a central plateau as well as more gentle foothills and coastal plains, the landscape supports vegetation ranging from tropical rainforest in the east to the extraordinary spiny forest of the arid south-west. However, this beautiful island has been systematically degraded and stripped bare over the centuries.

**Previous page:** Rosy periwinkle growing in Isalo National Park, south-west Madagascar. The sandstone rocks in this arid area also support many succulent plant species.

HE WORLD'S FOURTH LARGEST island, Madagascar is situated in the Indian Ocean, 400 kilometres from the south-eastern coast of Africa. Basic threats to the rich endemic flora and fauna result from the extreme poverty of the majority of the local Malagasy people, so tackling biodiversity conservation necessarily involves developing ways to improve the livelihoods of the rural poor. Almost 80 per cent of the country's inhabitants live in the countryside, where the farmers carry out subsistence agriculture, producing scarcely enough food to support their families; half of all children suffer from malnutrition.

Sadly, living conditions in Madagascar continue to decline across the island, with the arid south-west having the highest rural poverty rate. Environmental conditions continue to deteriorate and international greed for commodities such as rosewood (*Dalbergia*) and ebony (*Diospyros*) is wrecking habitats, as is the removal of tree fern trunks for use in international horticulture.

One traditional farming method on Madagascar is the 'tavy' system, which involves clearing the rainforest to plant maize, rice and manioc. After cropping, the tavy or rainforest clearing is abandoned and secondary vegetation known as 'savoka' usually takes over, with a mix of native and introduced species.

Scientists believe that prior to colonisation forest covered 90 per cent of Madagascar, but now less than 15 per cent of primary vegetation remains. Much of the island is covered by grassland that has developed since the earliest settlers arrived from Southeast Asia around 2,000 years ago. Each year, the grasslands are burned to produce new pasture for the zebu cattle that are the most valued possessions of the Malagasy farmers. A large herd of zebu is a sign of wealth, the value realized by the use of zebu horns in burial rituals. Other threats to the native flora of Madagascar include mining, tree felling, clearance of vegetation to plant commercial crops, and the targeted collection of attractive species such as orchids and succulents.

Eighty years ago, when Madagascar was governed by France, French botanists such as Perrier de la Bathie and Henri Humbert drew attention to the fragility of the island flora, noting the destruction of the rainforest and the potential transformation of the spiny succulent forest into a desert waste. Their work led to the creation of nature reserves and national parks that began in 1927. Humbert worked on the first comprehensive Flora of Madagascar, which was begun in 1936.

Palm-rich vegetation at Antrema, in the far north-west of Madagascar. The Antrema region is known for its diverse forest, grassland and coastal ecosystems and wealth of endemic species of wild fauna and flora. Increasing pressure on the natural vegetation is resulting from an influx of people displaced as result of climate change.

## MISSOURI AND MADAGASCAR'S FLORA

Baseline floristic information, describing which plant species occur where, based on documented herbarium specimens, is a crucial first step in planning the conservation and sustainable use of plant resources. Missouri Botanical Garden has been at the forefront of botanical exploration and plant conservation for many years and has helped to document the plants of Madagascar, China and Central America, as well as, closer to home, coordinating the Flora of North America project. Using the studies of taxonomic botanists as a basis, Missouri became involved in Madagascar in 1973 and this work continues to help to conserve the extraordinary diversity of plants and animals of this very special island. For more than 30 years, Missouri Botanical Garden staff have worked on the island, conducting botanical inventories, training local botanists and conservationists and, more recently, collaborating in community-based conservation.

Professor Peter Raven, the garden's director, explains why this long-term commitment is so important. 'Madagascar's flora is extraordinary, highly threatened and irreplaceable. We have to ensure that it is valued and understood both locally and globally. Once Missouri Botanical Garden became involved, I determined that we had to build up local expertise and stay the course.' Missouri now has a team of over 65 people in Madagascar dedicated to studying and conserving the unique and highly threatened flora. All but one of them are Malagasy, ensuring a continuing and full involvement in the preservation of the island's botanical riches.

The data collected by botanists is very important for conservation planning on the island. Missouri Botanical Garden has comprehensively documented and made available all names applied to Malagasy plants. The Madagascar Catalogue project is reviewing the taxonomic framework for each plant genus and is sorting out the species names together with their synonyms. A great deal of associated data is recorded for each species, including the threat status. So far, IUCN Red List categories and criteria have been applied to more than 3,000 of the endemic plant species. When Missouri began its work in Madagascar, the island's flora was estimated at about 9,000 species, but with many discoveries over the last few decades, it has become clear that the real number will exceed 13,000, with about 12,000 endemic to the island.

Madagascar's flora is extraordinary, highly threatened and irreplaceable

## MAPPING AND CONSERVATION

The Malagasy government has taken the initiative to triple the area included in national parks and reserves. In 2003, Marc Ravalomanana, then president of Madagascar, announced plans to protect a total of 6 million hectares, which is around 10 per cent of the entire country. Based on GIS data on plant diversity and analysis of the distribution of more than 1,200 endemic species, a map of 77 critical sites has been produced. A new, detailed, vegetation map of the country was published by Missouri Botanical Garden and the Royal Botanic Gardens, Kew in 2006. The government of Madagascar has now granted temporary protection status for 25 new sites within the national protected area system.

The highly restricted littoral forest of eastern Madagascar is a particular target for conservation action. A narrow belt of species-rich rainforest specially adapted to the sandy soils once ran parallel to the east coast of the island. Now it is estimated that over 90 per cent of this littoral forest has been lost. About 1,200 plant species, almost 10 per cent of Madagascar's flora, have been recorded from the remnant forests and about half these plants are found only in this habitat type. Botanists have identified the 15 most important sites for protection within this narrow strip of coastal forest, and they are included within the 77 national priority areas. Now Missouri staff are working to help communities in two areas of littoral forest to manage the forest resources sustainably while retaining the full range of plant diversity.

Flowers of *Intsia bijuga*, an important source of timber. Confined to coastal areas, this species has a widespread distribution but is now threatened in many countries because of its valuable wood.

## MAHABO FOREST

One such site is Mahabo Forest, an area of about 1,500 hectares with a particularly rich forest flora. Trees such as *Calophyllum, Tabernaemontana, Homalium, Diospyros, Intsia bijuga* and the endangered rosewood *Dalbergia chapelieri* grow with wild coffee plants, *Dypsis* palms, *Pandanus* and a profusion of orchids. Missouri botanists continue to find new species, such as a previously undescribed tree in the genus *Octolepis*. The forest is also home to a small population of the fruit-eating white-collared lemur *Eulemur albocollaris* – a species that is recorded as endangered. Scattered near the forest are small villages, where about 10,000 people traditionally make their living by tavy cultivation of rice and cassava. Fishing is also important and hunting takes place within the forest fragments. The forest also provides timber, fuel wood and a range of medicinal plants. Baskets woven from sedge are sold in the local markets.

Missouri has established a community-based conservation programme in the Mahabo forest region. Through workshops, sustainable use of forest resources is discussed and villagers are encouraged to protect rare species in their natural habitats. Project staff and local volunteers have planted out saplings of fast-growing eucalyptus species (*E. corymbosa* and *E. camaldulensis*) on the degraded grasslands between the villages and the forest to establish alternative sources of timber and fuelwood. Nurseries have been set up to grow saplings of native species for forest restoration. Fruit and spice trees are also being grown to provide food and income. One of the most exciting aspects of the Mahabo Forest Project has been training the local women to weave baskets in styles that will appeal to the international market. In the first year of production of the new designs the sale of baskets contributed over 10 per cent to the entire economy of Mahabo. Baskets are available in the Missouri Botanical Garden shop in St Louis, making an important link for garden visitors with the garden's work overseas.

*Pandanus biceps* (left) and *P. grallatus* (right). There are 75 species of pandanus native to the island. The leaves are used in basket making and the fruits for food.

## OTHER FOREST AREAS

Three hundred miles north of Mahabo, Missouri botanists conducted a floristic inventory of another area of littoral forest at Vohibola. Among the plants of this rich forest are five species of ebony, three species of *Dypsis* palm, two wild coffees, the cycad *Cycas thouarsii* and 15 orchid species including the spectacular *Angraecum sesquipedale*. This orchid has huge waxy yellowish flowers up to 12 centimetres across, with a delightful night-time fragrance. It is pollinated by a species of moth with a proboscis 22 centimetres long – as predicted by Charles Darwin who saw *Angraecum* in cultivation at the Royal Botanic Gardens, Kew in the 1860s. Another local plant is a tree named after the French colonial botanist, Humbert. *Humbertiodendron saboureaui* was last collected in 1949 and believed to be extinct until it was rediscovered in 2002. In 2001 a local NGO was appointed by the government to work with the local farmers in the Vohibola Forest area and help them to develop sustainable livelihoods. Missouri's baseline work helped to ensure that there is information to guide decisions about ecological aspects of sustainability. Fairchild Tropical Botanical Garden is now supporting this initiative.

Further inland, Missouri Botanical Garden is also working to preserve the plant resources of forests at Analalava and Ambalabe in collaboration with residents of neighbouring villages. The forest of Analalava is a rare rainforest fragment of about 200 hectares with over 350 plant species. In this small area there are 26 species of palms, 10 of which are found nowhere else in the wild. The forest also has a rich animal diversity with 5 lemur species, over 50 bird species, including the vulnerable crested ibis *Lophotibis cristata*, 33 species of reptile and 23 species of amphibian.

Analalava is seven kilometres from Foulpointe, a coastal tourist destination. As tourism has grown, so has the demand for timber for construction and felling of large trees has threatened the species-rich rainforest. Clearance for agriculture is the other main threat, despite the fact that as the forest is cleared, the water supply for irrigation of the local rice fields is adversely affected. Missouri scientists, almost all of them Malagasy citizens, are working with a local NGO, *Velonala*, to develop similar sustainable solutions as in Mahabo. A model nursery and five village-based nurseries have been established to produce native tree species and appropriate exotic species to grow for timber in the more severely degraded areas. Intensive permanent rice cultivation and a demonstration site for fish farming have also been developed. The project aims to show how the forest provides vital ecosystem services for local people, including surface and ground water essential for aquaculture and agriculture. Education is vitally important, enabling local farmers to become custodians of some of the rarest plants on earth. If the plans succeed, local people should also benefit from ecotourism and visitors will be able to learn about this unique environment.

Madagascar has a rich diversity of orchids. One of its more famous species is the spectacular *Angraecum sesquipedale.*

## SOHISIKA

Often, emblematic flagship species can be used to promote and support biodiversity conservation. Missouri is using this approach in the central highlands of Madagascar, where the critically endangered tree species *Schizolaena tampoketsana*, known locally as sohisika, is being used as a flagship for the conservation of fragmented remnant forests of Madagascar's High Plateau. Sohisika belongs to the family *Sarcolaenaceae*, one of Madagascar's six endemic plant families. This tree is known only from four tiny forest fragments on the plateau known as Tampoketsa of Ankazobe and as scattered individuals in nearby highly degraded grassland. Only 160 individuals survive in the wild, and these trees remain highly threatened by selective felling and annual burning.

Missouri launched a rescue mission for sohisika in 2004, aiming to ensure that the species survives in its natural habitat and that its conservation draws attention to the critical status of the forest remnants of the central highlands. Priority attention is being given to securing a population of sohisika in a 33-hectare forest fragment of Ankafobe, one of the largest remaining forest patches in the area, which was found to contain approximately half of the remaining sohisika trees.

The forest has been damaged by timber felling and the ubiquitous burning – threats which must be halted if sohisika is to survive. Local people have to be involved in developing the solutions and they are now managing the site with guidance from Missouri staff. Under the new management regime, the forest is showing signs of natural regeneration. In the more degraded areas, thousands of native trees of 25 species, including sohisika, have been planted.

The sohisika project is vitally important as a pilot for conservation success on an island that has seen much ecological devastation and species loss. Ankafobe is only 15 kilometres from the Réserve Spéciale d'Ambohitantely. This 5,600-hectare reserve is the largest natural forest on the central highlands and provides a habitat for all kinds of animals. If ecological restoration on the ambitious scale envisaged by the Missouri botanists is to be successful, this protected area will serve as a 'reference' ecosystem, showing the species composition and forest structure to aim for, to the extent possible given the degraded soils. It will also be a reservoir of plant material for use in future restoration.

## FUTURE CONSERVATION PROJECTS

Missouri Botanical Garden's Malagasy staff are planning to build on their successes and develop further conservation and sustainable projects throughout Madagascar, paying attention to important areas of plant diversity and endemism that have so far slipped through the conservation net, possibly because of the absence of charismatic mammal species. Having locally trained people is one of the keys to success, but continuing support from overseas will also be crucial. In Madagascar, Missouri Botanical Garden is leading a range of training activities for the local Malagasy staff designed to build up the practical botanical skills and experience needed for conservation planning. This programme provides training in the basic botanical fieldwork which is important in the frontline of research and conservation action. University students are also trained in specialist botanical skills and a few Malagasy research botanists are able to complete PhD studies in St Louis.

As conservation action on the ground expands, Missouri staff are planning to work in the Ibity Massif, a mountain of quartzite near Antsirabe. This area is very important for succulent plant conservation with endemics such as *Aloe trachyticola*, and the vulnerable *Pachypodium brevicaule*, *Pachypodium eburneum*, which may now be extinct in the wild. Interesting and threatened orchids, such as the vulnerable *Angraecum sororium* are also present. The presence of these fascinating plants has acted as a magnet for unscrupulous plant collectors and there has been a major problem in the past with collecting for the international market. BGCI is joining conservation efforts in this area as part of a broader initiative to learn from and promote links between botanic gardens and local communities.

**Left:** *Adansonia rubrostipa*, one of the six endemic species of baobab growing in Madagascar.

**Right:** *Pachypodium mikea*. This tree-like succulent was described for the first time in 2005.

## ILE SAINTE MARIE

Another area where Missouri is focussing its attention is the idyllic Ile Sainte Marie, a tiny island 14 kilometres off the east coast of Madagascar that is rich in endemic plant species. From 1685, Ile Ste Marie was the centre of piracy in the Indian Ocean. From the shelter of the island's bays, legendary pirates, such as William Kidd, plundered ships returning from the East Indies laden with treasures. Remains of pirate vessels still lie off the coast and the pirates' cemetery at Baie de Forbans is a reminder of this peaceful island's turbulent past.

In 2008 Cyclone Yvan passed directly over Ile Sainte Marie. Heavy rains and winds exceeding 240 kilometres per hour caused tremendous destruction and human tragedy. The fragments of natural rainforest were badly affected. Even prior to the cyclone, some of the endemic species were extremely rare; a recent inventory revealed only nine individuals of *Mantalania longipedunculata*. According to Missouri Botanical Garden's Chris Birkinshaw, the impact of post-cyclone human activities may be very bad for the local flora: 'Typically, following cyclones the impoverished population are forced to exploit 'communal' timber resources to rebuild their houses or to make money to buy food to replace the harvest destroyed by the wind and rain. Without careful planning and support for local people, the impact of Cyclone Yvan and the associated increase in exploitation of timber resources by local people may really push some of Ile Sainte Marie's endemic plants to the very brink of extinction or beyond.' Missouri is working to locate remaining populations of significant plants to ensure their *in situ* and *ex situ* conservation.

## DEVELOPING LOCAL EXPERTISE

Missouri botanists work closely with Madagascar's main botanic garden, the Parc Botanique et Zoologique de Tsimbazaza, which was established in the capital city of Antananarivo with support from French botanists in 1925. The garden is located by the artificial lake of Tsimbazaza, created by King Radama in 1815. During the 19th century, the lake was a site for promenading and for ceremonial sacrifices of zebu cattle. The garden is now developing its expertise in botanical conservation and is also a good place to see Malagasy endemic orchids, such as the spectacular *Angraecum sesquipedale*. Other species of *Angraecum*, for example *Angraecum longicalcar*, may already be extinct in the wild but fortunately survive in conservation collections.

Botanic gardens alone cannot solve all Madagascar's conservation problems, but they can highlight the plants and plant-rich places that particularly need protection, and they can supply the horticultural knowledge and skills vital in repairing damaged ecosystems. Missouri's conservation work in Madagascar

A nursery for the production of native tree species at Agnalazahe Forest. Replanting using native trees is important in ecological restoration and supports carbon sequestration.

# SAVING SUCCULENTS

Madagascar's succulent plant flora is truly remarkable, with over 600 species, most of which are endemic. Many of these species are propagated readily in cultivation and can be seen in botanic gardens around the world. Within Madagascar, the spiny desert of the south-west is perhaps the most amazing habitat of all. In a small coastal region about 50 kilometres wide, extraordinary dryland vegetation made up of succulent *Euphorbia* species and plants of an endemic family, the Didiereaceae, grow on thin skeletal soils. The genera *Didierea, Alluaudia, Alluaudiopsis* and *Decaryia* are all confined to the south and south-west of the island. All species of these genera are threatened with extinction in the wild because of the vulnerability of their habitat, which is being cleared by local farmers eking out a subsistence living by growing maize. Charcoal production also threatens the spiny forest.

The Arboretum d'Antsokay in Tulear is a remarkable botanic garden dedicated to the display and conservation of Madagascar's unique succulent plant flora. Founded by Hermann Petignat and his wife Simone in 1980, the garden is now run by their son Andry Petignat. About 540 species grow in the collection, including the Didiereaceae and the critically endangered *Aloe suzannae*. Building up the resources of this arboretum will be very important for conservation in the Tulear region – the poorest part of Madagascar. Andry is currently striving to develop education programmes at the garden.

*Aloe vaombe* growing wild. This species is one of many Madagascan succulents now seen in gardens around the world.

has developed over the past 30 years and is undertaken in partnership with other gardens around the world and the Malagasy botanic gardens themselves.

George Schatz, one of Missouri's experts on the flora of Madagascar and a world leader in conservation assessment, is cautiously optimistic about the future. 'I have been visiting Madagascar for 20 years and remain fascinated by the extraordinary plant diversity. My hope is that we can join up all the small projects that Missouri has initiated to form a network of community-run projects across the island. Community action has to be the key to lasting conservation success.'

South Africa is a country of botanical superlatives, truly a plant-lover's paradise. It has an extremely rich and diverse flora with 19,600 native plant species, many of which are endemic. Many garden plants grown in Europe and elsewhere in the world have their origins here: ericas, gazanias, osteospermums, pelargoniums, mesembryanthemums, proteas, and bulbs of amaryllis, moraea, nerine and gladiolus are just some of the ornamental plants that have been brought into garden cultivation from South Africa. Half the world's succulent plants grow wild in South Africa and the diversity of bulbous plants is stunning. A network of botanic gardens across the country plays a vital role in the conservation of this rich natural heritage.

The Cape Floristic Region, one of the world's six 'floral kingdoms', is entirely contained in an area of 89,000 square kilometres in the south-western part of South Africa. This unique area is home to about 8,550 species of flowering plants and ferns, of which over 70 per cent are endemic to this part of the country. Species-rich heath or 'fynbos' (from the Dutch fijn bosch which means 'fine bush') is the dominant vegetation type. The Cape Floristic Region is recognized as a World Heritage Site and as one of the world's outstanding priorities for plant conservation. Kirstenbosch National Botanical Garden, on the outskirts of Cape Town, is included within the World Heritage designation as part of the Table Mountain National Park – the first botanical garden in the world to be included within a natural World Heritage Site.

Kirstenbosch National Botanical Garden is an internationally renowned attraction for everyone interested in growing and studying plants. The garden occupies a dramatic setting at the foot of Table Mountain and includes over 490 hectares of natural vegetation. The cultivated area, of 36 hectares, primarily grows indigenous South African plants. The garden was established in 1913 on land bequeathed to the nation by Cecil John Rhodes, and from the outset, the

SOUTH AFRICA

*Erica verticillata* became extinct in the wild and is now cultivated in botanic gardens and being reintroduced to its native habitats.

first Director of the Garden, Professor Harold Pearson, recognized the importance of conservation as one of the key roles of Kirstenbosch, alongside botanical research and education.

Kirstenbosch aims to 'promote the sustainable use, conservation, appreciation and enjoyment of the exceptionally rich plant life of South Africa, for the benefit of all its people'. It is part of a countrywide network of nine National Botanical Gardens managed by the South African National Biodiversity Institute (SANBI). SANBI was created by law in 2004 to look after the rich biodiversity of the country. Through this arrangement, plant conservation is truly integrated in South Africa, with all aspects of *ex situ* and *in situ* management of plants developed under the same conservation umbrella.

## A NATIONAL NETWORK

Conservation of South Africa's native plants is vitally important and urgent. Thirteen per cent of South Africa's native plants are regarded as threatened with extinction, with two per cent classified as Critically Endangered. Agricultural development has been the predominant threat to plant diversity, together with

Kirstenbosch National Botanical Garden has a magnificent setting against the eastern slopes of Cape Town's Table Mountain.

urbanization, mining, the unsustainable collection of valuable species and a huge problem presented by invasive species introduced from Australia and other parts of the world.

The National Botanical Gardens are located in six of South Africa's nine provinces and include a combined area of 7,500 hectares of natural vegetation within their boundaries. With temperate, Mediterranean, semi-arid, and subtropical to tropical climates represented, they are collectively able to help protect representative plants and plant communities from different ecosystems.

SOUTH AFRICA

Six of southern Africa's seven biome units are maintained: forest, fynbos, grassland, savannah, Nama Karoo and Succulent Karoo. The only biome not represented is desert, which is not really included within South Africa's boundaries, the nearest being the Namib Desert of neighbouring Namibia.

The gardens contain around 8,500 indigenous plant species, over one-third of South Africa's stunning native flora. They hold 17 per cent of South Africa's estimated 2,301 threatened plants in *ex situ* collections – genuinely acting as botanical arks. Threatened species from the ornamental families Proteaceae,

Amaryllidaceae, Aloaceae and Iridaceae are particularly well represented. Plants are held in living collections, which are usually part of the garden displays, or stored in seed banks. The gardens aim to ensure that duplicate material of very rare species is held by across the network to minimize the risk of *ex situ* stock being lost. In some cases, botanic gardens overseas share the responsibility for maintaining *ex situ* material of South Africa's precious plants, particularly European gardens that have long collaborated in botanical exploration and research in Southern Africa.

The conservation influence of the National Botanical Gardens extends way beyond the boundaries of the gardens themselves. Locally, for example, Kirstenbosch is responsible for the management of Edith Stephens Wetland Park on the Cape Flats and the Tienie Versfeld Reserve near Darling in the Western Cape.

At a national level, the gardens are very closely involved with the Botanical Society of South Africa. The society was established in 1913 specifically to support the development of Kirstenbosch and now works with all the gardens in the network. Members of the Botanical Society act as the 'friends' of the gardens and support both garden-based and *in situ* conservation action. In recent years, South Africa's botanic gardens have broadened their activities with urban communities, notably with programmes that use native plants to 'green' disadvantaged schools in township areas around the gardens in the major cities of Cape Town, Bloemfontein, Nelspruit, Pretoria and Johannesburg. This is a vitally important way to engage more people in celebrating and caring for native species.

Regionally, the gardens have helped to support plant conservation throughout Southern Africa and internationally the gardens provide models for integrated conservation by botanic gardens. The overall policy for plant conservation in the SANBI botanic garden network is set out in South Africa's

The King Protea, *Protea cynaroides*. Kirstenbosch displays a wide variety of plants of the Cape Floral Kingdom. The best time to see proteas in flower at the garden is between August and October.

formal response to the Global Strategy for Plant Conservation. Christopher Willis, who is in charge of the network, worked with many contributors from South Africa's conservation community to compile this national report. He considers that conservation of South Africa's plant diversity is 'an awesome responsibility. We have so many rare plants packed into a small area and the threats increase each year. Fortunately, willingness and the skills to protect our unique flora are also developing rapidly and we have excellent collaborative networks. It is exciting in South Africa that botanic gardens blur the distinctions between *ex situ* and *in situ* conservation.'

The SANBI headquarters are based in the Pretoria. Compared to Kirstenbosch, this is a relatively young garden, formally established in October 1958. The other South African gardens are Free State, Hantam, Harold Porter, Karoo Desert, Lowveld, KwaZulu-Natal and Walter Sisulu National Botanical Gardens. The Hantam National Botanical Garden, established in August 2007 on the outskirts of the small town of Nieuwoudtville in the Northern Cape, comprises over 6,300 hectares of land on the Bokkeveld Plateau, and is world

A carpet of *Brunsvigia bosmaniae* at Hantam National Botanic Garden. This spectacular autumn display occurs after late summer rain.

renowned for its incredible diversity of bulbous plants. Some 40 per cent of the flora comprises bulbs that create spectacular displays in autumn and spring each year. The garden also has large natural patches of renosterveld, fynbos and Succulent Karoo vegetation.

This new national botanical garden comprises an important conservation area that will be used by SANBI to promote the conservation of the area's unique biodiversity through nature-based tourism, environmental education opportunities and long-term ecological research into this botanical hotspot of global significance.

### KAROO DESERT NBG

The Karoo Desert National Botanical Garden is located at the foot of the Brandwacht Mountains near Worcester, 120 kilometres north of Cape Town, in an area that experiences an annual rainfall of only 250 millimetres. The Karoo and Namib Deserts are simply remarkable for their plant diversity and the number of endemic species. The arera is known as the Karoo-Namib Centre of Endemism or the Succulent Karoo. This broad arid region, defined by its plant diversity, is one of 25 global biodiversity hotspots identified by Conservation International and it is the only one that is entirely arid. The Karoo Desert National Botanical Garden is one of only a handful of botanic gardens in this area and is considered to be the only truly succulent botanic garden in Africa.

Approximately 11 hectares of the garden are cultivated, while the remaining 144 hectares are kept as a flora reserve, with nature trails weaving through the native karroid vegetation. Around 2,500 succulent plant species are grown within the garden, roughly half of all Southern Africa's succulent diversity, including important reference and conservation collections of *Haworthia, Conophytum*, asclepiads and succulent euphorbias and many species that are very rare and under threat in the wild. The plantings of the garden reflect the floras of different areas of the Karoo-Namib Centre of Endemism, such as the Richtersveld, the Knersvlakte, Tanqua Karoo and Little Karoo. Spectacular plants that can be seen here include the conservation flagship species *Pachypodium namaquanum* and *Aloe pillansii*, extraordinary tree-like succulents.

*Didymaotus lapidiformis* is considered to be Vulnerable in the wild due to habitat loss and over-collecting for horticulture. It is being conserved by the Karoo Desert National Botanic Garden.

*Aloe pillansii* has declined dramatically to fewer than 200 individuals in the wild and is now Critically Endangered. The few remaining wild plants provide an important source of shelter, nectar, food and moisture, particularly for birds, but mature plants are dying, partly as a result of leaf scale disease, and it seems that the species is not regenerating in the wild. Baboons and porcupines gnaw the stems and grazing by goats is also thought to be a problem. Collectors of rare plants have also contributed to the decline of this tree aloe.

*Aloe dichotoma*, commonly known as the quiver tree, growing at the Karoo Desert National Botanic Garden. The range of this species has changed significantly as a result of climate change.

Another attractive succulent species being conserved by the garden is *Didymaotus lapidiformis,* a tiny 'stone plant' that is well camouflaged in the rocky desert terrain. Also a candidate for conservation is *Haworthia maxima,* which was rescued from a development site in Worcester and, following propagation in the botanic garden, will be restored to the wild once the building development is completed.

## WALTER SISULU NBG

The Walter Sisulu National Botanical Garden is situated approximately 25 kilometres west of Johannesburg in an area of vegetation known as the rocky highveld grassland, a mosaic of grassland and savannah with trees such as *Acacia* and *Combretum molle.* The various habitats within the grounds of the garden support over 600 native plant species and over 240 bird species, including a breeding pair of majestic Verreaux's eagles that nest on the cliffs alongside the garden's scenic waterfall. The garden has a cycad collection and also a succulent rockery. Threatened species that are being conserved here include *Aloe albida* and *Aloe peglerae. Aloe albida* is a dwarf grass aloe that occurs in montane habitats in the Mpumalanga Province in the north-eastern part of South Africa and also in Swaziland. This attractive plant produces a single inflorescence with small white flowers and has suffered in the wild as a result of collecting pressures. Plants are being grown at the garden from seed collected in the wild, and there are plans to reintroduce the species by sowing seed in rocky habitats where small pockets of still survive in the wild.

*Aloe peglerae,* a small chunky stemless aloe with coral-coloured flowers, occurs on north-facing slopes of the Magaliesburg and Witwatersrand in the North West and Gauteng Provinces of South Africa. Illegal collection poses one of the greatest threats to this species. Plants are being cultivated at the Walter Sisulu garden to sell to the public in the hope that this may take the pressure off wild populations.

## CYCADS

Cycads are an extraordinary group of plants related to the conifers that flourished in the time of the dinosaurs. Today, half the world's cycad species are threatened with extinction. South Africa has 35 native species, 11 of which are Critically Endangered and all of which are in cultivation in the SANBI botanic garden network. Cycads are one of the key groups of plants displayed and

# A 'WONDER CURE FOR OBESITY' THREATENED BY EXTINCTION

South Africa has about 3,500 plants with medicinal uses. One of the most remarkable plants that has recently come to international prominence is the succulent species *Hoodia gordonii*. This fascinating plant was discovered in December 1778 by the botanist Francis Masson, the first plant collector employed by the Royal Botanic Gardens, Kew. It is a spiny succulent shrub with large pink or red petunia-shaped flowers that smell similar to rotten meat and attract flies (which are the pollinators). Confined to arid areas of Namibia and South Africa, for centuries the nomadic San people have used this plant as an appetite suppressant to stave off thirst and hunger pains during hunting expeditions. In recent times *Hoodia gordonii* has been promoted worldwide as a natural slimming aid. This interest has placed huge pressures on wild populations of *Hoodia gordonii* and its close relatives, which look superficially similar.

Conservation of *Hoodia gordonii* requires a fully integrated approach and international collaboration. Since 2005 the entire genus has been listed on Appendix II of CITES, with provisions in place to encourage trade in sustainably harvested plants. This is designed to act as an incentive for conserving the species as a valuable source of income for local landowners and collectors. But this policy is challenging to enforce and illegal collecting remains a problem. DNA barcoding has been introduced to assist with tracing the source of the material and also with quality control. Cultivation is also likely to play an important role. *Hoodia gordonii* is in cultivation at the Karoo Desert National Botanical Garden and is also cultivated by private growers. SANBI is helping provincial conservation authorities develop strategies for the future survival of this species.

*Hoodia gordonii* growing in the Tanqua Karoo, protected from goats by wire netting. Cultivation of this species offers an important new source of income for farmers in South Africa.

Hand pollination helps to ensure seed production, aiding in the long-term survival of cycads such as this species, *Encephalartos princeps*, which is threatened in the wild by habitat loss, collection for horticulture and invasive species.

conserved at Kirstenbosch, reflecting a historic interest of the garden since its establishment nearly a century ago. Some of the earliest plantings at the garden were of cycads such as *Encephalartos latifrons*, a native species that is now Critically Endangered in the wild following a decline as a result of habitat loss and over-collecting for the horticultural trade. Fewer than 60 plants now survive in the wild and individuals are too far apart for successful pollination to occur. The remaining plants have all been fitted with microchips to ensure that illegally collected plants can be identified if they are confiscated by enforcement agencies. The 40 plants, including seedlings, in cultivation at Kirstenbosch provide an important source of material for reintroduction to the wild as they have been shown to have genetic diversity that will be appropriate for reinforcing the wild populations.

A wide variety of *Encephalartos* species can be seen growing in the Cycad Amphitheatre at Kirstenbosch, including a caged specimen of *E. woodii*, which is now extinct in the wild and exists only in botanic gardens. Only one specimen of this cycad was ever known in the wild – a solitary male plant found in Ngoya Forest in South Africa in 1895. Three of its four main stems were collected and have been the source of all the material now grown in botanic gardens around the world. Kew received a stem of *E. woodii* in 1899 and the plant can still be seen growing in the Temperate House.

## ECOLOGICAL RESTORATION

Research and conservation action for individual threatened plant species is extremely important if we are to reverse the drift towards extinction. Work undertaken by South Africa's botanic garden network is taking this a stage further by looking at ecological restoration of whole plant communities. Fynbos vegetation is fire-adapted, which means that individual species and plant associations are dependent on burning for their long-term survival. The Western Cape Lowlands within the Table Mountain National Park have a special type of fynbos known as Cape Flats Sand Fynbos. The plants that grow there include a mix of bulbs, such as the beautiful peacock iris (*Moraea aristata*), reeds in the Restionaceae (an endemic South African family), and shrubs in the Proteaceae including *Leucadendron* and *Leucospermum* species. The spread of invasive trees and grasses is a major threat to the fragile and fragmented Cape Flats Sand Fynbos and a range of other problems include urban development and illegal harvesting of thatching reed. Minimizing the threats to the habitat is essential before restoration can take place and this involves close cooperation with the people who live in the area.

Carefully controlled burning is used to clear the ground at the start of the restoration process for the Cape Flats Sand Fynbos. This clears invasive species, releases nutrients into the soil and allows some of the native plants, especially bulbs, to re-establish. Other depleted species need to be reintroduced. Seeds of endangered species are treated with smoke from woody fynbos vegetation under controlled conditions at Kirstenbosch to aid germination. They are planted in small bags and grown for a year before being planted out into the wild. A number of species have been reintroduced in this way, including the attractive heath, *Erica verticillata*. This species was lost in the wild by the early 20th century as a result of agriculture and urban development. In the 1980s, a handful of plants were found to have survived precariously in a garden in Pretoria, in Kirstenbosch and at the Royal Botanic Gardens, Kew. All these plants were successfully propagated from cuttings and reintroduced into horticulture. Now plants are being reintroduced into the wild – a century after the presumed

*Moraea aristata* only survives in a small area of native habitat in Cape Town but has become invasive in other parts of the world.

extinction of this attractive shrub. According to Carly Cowell, one of the researchers at Kirstenbosch, 'It is hardly an exaggeration to say that the fate of a species lies in recognition that conservation and restoration are just different ways of doing the same dance. Propagating and displaying threatened plants in botanic gardens is vitally important as is long-term storage in seed banks, but true conservation involves ensuring that species are returned to, and survive in, their natural habitats.'

Challenges to restoring a fynbos habitat include ensuring that plants raised in the garden can adapt to low nutrient levels in the soil and survive predation by the abundant small mammals, such as mole rats and gerbils. Kirstenbosch's programme is helping to show people how to tackle these practical problems and, over time, create a vegetation structure similar to

*Agapanthus praecox,* popular in gardens throughout the world, is native to South Africa. Here it is growing wild at Mossel Bay, Southern Cape.

*Serruria furcellata,* a plant of the Proteaceae family, is Critically Endangered and reliant on careful conservation management to ensure its survival.

natural fynbos. Once this is in place, other indigenous plants and animals have the may colonize from nearby. Eventually a more complete natural veld may develop within Cape Town, where remaining green areas such as rivers, wetlands and green belts can be used as corridors for natural fauna and flora to proliferate.

Botanists are also working at the Plattekloof Natural Heritage Site in the Western Cape Lowlands, where a narrow corridor of land owned by South Africa's power supply company is surrounded by densely populated urban areas. Kirstenbosch scientists are restoring a number of endangered plant species, including *Serruria aemula,* which is reduced to about 1,000 plants in the wild. It is displayed at Kirstenbosch and has been propagated from cuttings for successful re-establishment at its native site. Other plants that are being restored are *Serruria trilopha, Leucadendron levisanus* and *Diastella proteoides,* a low-growing shrub with dainty pink flowers.

The specialized fynbos vegetation within the urban area of Cape Town may be particularly threatened, but throughout the Cape Floristic Region, fynbos is in trouble – as are many other species-rich vegetation types. Plant conservationists in South Africa really are engaged in a race against time. The National Botanic Gardens network provides a refuge for rare and endangered plants from fynbos, arid lands and the different habitats, enabling conservation options to be kept open for the future. Looking ahead, closer integration of *in situ* and *ex situ* conservation is a key goal for the network. Monitoring and evaluation of the conservation role and status of *ex situ* collections is vitally important. At a time of rapid climate change, the stored plant material will become increasingly valuable, as will knowledge of how to cultivate rare and threatened plants and how to re-establish species in the wild.

Botanic gardens in Australia have a strong commitment to the conservation of the country's native flora. When the first botanic gardens were established in the 19th century their emphasis was on recreating  temperate floral displays reminiscent of the gardens of Europe. Over time the focus has shifted squarely to studying and looking after Australia's native plant diversity. As global climate change causes uncertain weather conditions and increasing periods of drought, scientists are looking for ways to both conserve and utilize the native flora. Australia's native vegetation varies from tropical rainforest, through woodlands, spinifex grasslands and Mediterranean heaths to sandy desert with an ephemeral flora.  The country's plants have been well documented and the national list of threatened plants consists of about 1,400 species. The botanic garden network is well-placed to help protect them.

**Previous page:** Scientists at Kings Park and Botanic Garden in Perth, Australia, are world leaders in orchid conservation.

South-west Australia, the southern part of the state of Western Australia, is a global biodiversity hotspot. The original vegetation of this part of the country was predominantly eucalyptus woodland, eucalyptus dominated mallee shrubland and various kinds of heath. South-west Australia has a third of Australia's flora and more than half of these plants are endemic to the region. There are 246 different species of eucalyptus in the area varying from trees such as the majestic jarrah (*E. marginata*) and karri (*E. diversicolor*) to small mallee shrubs such as *E. preissiana*. Australia's rarest tree, Eucalyptus phylacis, known only from a single clonal population, occurs within south-west Australia. Many species of the protea family are found here. Banksia, named after the botanist Sir Joseph Banks, has its greatest diversity in this species-rich corner of Australia.

But this incredible corner of Australia is in trouble. Many of the species are naturally rare with a small population size. Extensive clearance of the vegetation for agriculture since the arrival of the Europeans has been a major threat to these fragile species and mining is an additional threat. Throughout the wheatbelt region of Western Australia over 90 per cent of the natural vegetation has been cleared. Clearing the vegetation has led to rising groundwater levels

and salinization. Competition with introduced weeds is an ongoing problem and a particularly pernicious threat has been the arrival and spread of one of the world's most damaging invasive species, the root-rot fungus *Phytophthora cinnamomi*. Dieback disease caused by this fungus is a particular threat to banksias and grevilleas.

*Banksia brownii* is one of the very many plant species of south-west Australia that is endangered in the wild. Confined to widely separated sites in the Stirling Ranges and north-east of Albany, the main threat to this attractive ornamental shrub is infection by *Phytophthora cinnamomi*. Fortunately the species is in cultivation at a number of botanic gardens around the world and is one of 40 banksia species that can be seen growing at Kings Park and Botanic Garden in Perth, the capital city of Western Australia. Reintroduction may be possible if the threat from the fungus can be tackled.

## KINGS PARK AND BOTANIC GARDEN

Kings Park and Botanic Garden is a popular attraction in Perth and is also an internationally respected centre of plant conservation expertise. One of the

Banksia heathland at the Moore River National Park, Western Australia. Many species of banksia are now under threat in the wild as their habitat declines.

eight capital city gardens, Kings Park gives priority to conserving plants of the South-west Australia Global Biodiversity Hotspot. The site was once known as Perth Park, created in 1872 on land that had previously been an important ceremonial and spiritual dreaming place for people of the Nyoongar nation. The name of the park was changed to Kings Park to mark the accession of King Edward VII in 1901.

The initial idea for the parkland was for an English style park with avenues of shady trees. The climate and soils were not, however, conducive to this romantic vision and the introduced trees such as English oaks failed to thrive. Most of the site remained as bushland. In 1959 a State Botanic Garden was established in Kings Park complementing the botanic gardens in other Australian states which were developed in the colonial period. From the outset there was a commitment to cultivating the native flora of the south-west of the state. Natural bushland remains in the garden today, enhanced with plantings of the rich local flora. In this way the garden is able to display the natural plant wealth of Western Australia to the five million visitors who enjoy the garden each year.

Professor Kingsley Dixon, Director of Science at the garden, has had a life-long interest in the local flora. 'I grew up within easy reach of bushland areas in the country and outskirts of Perth. This probably explains why my interest in native plants began at an early age with forays into

Conserving the plants of Western Australia is important to support indigenous birdlife such as this white-cheeked honeyeater.

surrounding bushland in search of orchids and other charismatic plants. This combined with a deep interest in plant propagation and cultivation during my primary school years which has never left me and so I have been very fortunate to develop a career based on those early hobbies.' Having researched the ecology and physiology of Australian native plants and ecosystems for over twenty years, Kingsley now leads a research team of over 40 research staff and postgraduate students specializing in seed ecology and biology, propagation science, germplasm storage, conservation genetics and restoration ecology.

## RECOVERY PROGRAMMES

One of the species that Dixon and his team are helping to protect is the extraordinary Western Australian underground orchid, *Rhizanthella gardneri.*

This is one of only two orchids known to live their entire life cycle under the surface of the soil. *Rhizanthella gardneri* has no leaves or chlorophyll and so cannot produce its own food. Instead it derives nutrients from a symbiotic relationship with the broom honey-myrtle (*Melaleuca uncinata*) and specific mycorrhizal fungi. The broom honey-myrtle grows in remnant fragments of specialized mallee heath within the wheatbelt.

Flowering of *Rhizanthella gardneri* generally begins in the Australian autumn. Around the end of May each plant produces up to 100 small cream to reddish coloured flowers, surrounded by large, cream or pinkish-cream bracts. These bracts give the orchid a tulip-like appearance and form a small opening at the soil surface. A layer of leaf and bark litter covers this opening. The flowers grow from a horizontal rhizome below the ground, which, like the rest of the plant, is succulent and produces a formalin-like smell when cut. After pollination by termites or small midges, each flower produces a berry-like fleshy fruit containing 20 to 150 seeds. This type of fruit is unique among the Western Australian orchids as all other species produce a dehiscing pod from which thousands of minute seeds are dispersed by the wind. In contrast the seeds of *Rhizanthella gardneri* are thought to be dispersed after being eaten by small native marsupials.

*Rhizanthella gardneri* is Critically Endangered with a total population size of less than 100 mature individuals. Once known only from the Corrigin–Babakin area in the central wheatbelt where it was discovered 80 years ago, Dixon and other members of the local orchid society found new populations in the Munglinup–Oldfield River area some 260 kilometres away in the early 1980s. Extensive surveys elsewhere in the region have not revealed any further plants. Increasing drought has placed a strain on the habitat of the underground orchid and many of the honey-myrtle bushes are dying of old age, which is leading to a more open habitat with greatly increased light levels and altered soil organic matter. In addition to fragmentation and changes to the habitat, damage by people visiting the sites to find the orchid have put additional strain on the tiny populations.

Fortunately, a recovery plan has been developed for the species. As part of the plan, two reserves have been set up at Babakin giving legal protection to wild populations. Under Dixon's supervision, Kings Park and Botanic Garden is looking after seed collections of *Rhizanthella gardneri* and the associated *Melaleuca uncinata*. Mycorrhizal fungi have been isolated from the orchid

The extraordinary underground orchid, *Rhizanthella gardneri*, has a tiny population in the wild and is being conserved through joint efforts scientists, orchid enthusiasts and local farmers.

rhizome at Babakin and the garden has developed methods for the production of mycorrhizal inoculum suitable for glasshouse and field studies. Trials on germination have begun and there is hope that this extraordinary orchid can be saved. 'Conserving this iconic orchid, one of the world's most bizarre plants, with its complex ecology is a challenge,' says Dixon. 'Fortunately there are enough people in the government and scientific community, together with orchid enthusiasts and local farmers, who care about a successful outcome for the Western Australian underground orchid. It may not have known utility value but is one of those plants with a fascinating story and unique appeal.'

Another wheatbelt species being conserved by Kings Park and Botanic Garden is Bancroft's symonanthus (*Symonanthus bancroftii*) one of the rarest species in Western Australia. The last known botanical collection of this shrub was in the 1940s and until recently it was thought to be extinct. However, the remarkable discovery of a single male plant on a road-verge remnant in 1997 led to a determined search. Fortunately the next plant discovered was a single female growing beside a railway track. Both sites were subsequently protected with fencing but this could not significantly reduce the variety of risks facing the fragile individuals. As the male and female were widely separated botanists thought it unlikely that natural pollination could occur or would occur quickly enough to establish a soil seed bank before the plants died. Urgent *ex situ* propagation was the only option. Taking traditional cuttings was considered to be too risky because the two plants were very small so tissue culture was chosen for rapid propagation.

Over the past decade Dr Eric Bunn, head of conservation biotechnology and his collaborators at King's Park and Botanic Garden have successfully set *Symonanthus bancroftii* on the road to recovery. Initially, with permission from the state conservation authority, a small amount of shoot material was collected from the male plant and propagated using tissue culture techniques. Similar techniques were then applied to material from the female plant. After a few years both male and female micropropagated plants flowered and the female flowers were successfully pollinated artificially using pollen from the male plants. This resulted in the production of the first seed of this species to be seen in over 50 years. Over 2,400 male and female plants produced from the micropropagation programme were planted in field sites from 2002–2006. Survival was low overall as many of these years experienced drought

*ex situ* conservation and restoration of alternative sites will become more and more important

conditions and artificial watering from nearby depleted farm dams proved to be very difficult. Nevertheless some plants survived and were pollinated by native insects so that seed was produced naturally in the wild. It is now thought that there are enough plants to sustain a stable reintroduced population capable of providing enough seed for further restoration efforts.

This remarkable project has provided an invaluable opportunity to study the effects of extreme loss of genetic provenance on a wild plant species and the capacity of a species to recover from the very brink of extinction. Biologists know that some plant species have undergone quite extreme genetic 'bottlenecks' in the past and were able to recover naturally. The reintroduced populations of *S. bancroftii* represent a unique *in situ* 'laboratory' for the study of recovery potential. As with all the conservation efforts of King's Park and Botanic Garden, the continued survival of *Symonanthus* relies on partnership between conservation agencies and volunteer community groups working with scientists and research students to monitor the plants in the wild.

Dixon and his colleagues accept that for many of the world's threatened plant species, repatriation to their natural habitat is no longer an option. Of the 2,814 species of conservation concern in Western Australia, for example, at least half exist in highly disturbed habitats where salinity, invasive species adn diseases mean that the ecological conditions needed to support long-term sustainable populations are not in place. *Ex situ* conservation and restoration of alternative sites are set to become increasingly important. Such is the case for *Grevillea pythara*, now reduced to a single genetic individual in the wheatbelt of south-western Australia. Threatened by salinization, invasive weeds, agricultural chemicals and grazing by rabbits, the only future for conservation of this plant in natural conditions will be restoration of nearby landscapes and assisted migration to new safe sites. *Ex situ* collections at Kings Park and Botanic Garden are available to use in this recovery process.

# RESTORING DEGRADED HABITATS

Around the world ecologists and conservationists are working to match the pace of landscape alterations with the ecosystem restoration and ecosystem reinstatement. Botanic gardens are supporting ecological restoration using material from seed banks and specialized knowledge of plant diversity, the structure of habitats and cultivation techniques. The extensive nursery and experimental section at Kings Park has researched the cultivation of many

Restoration of mine sites is supported by research at Kings Park. Conservation partnerships with mining and quarrying companies are increasingly important in Australia.

different native plants, from orchids to eucalypts. Halophytes, or salt-tolerant plants, some of which have economic importance for livestock grazing, are studied with particular interest. With increasing salinization of natural habitats, understanding how these species thrive is likely to become more and more relevant. Building on the conservation and cultivation techniques for single species, Kings Park and Botanic Garden has a major commitment to the restoration of degraded natural habitats in Western Australia using indigenous plant species. In this work of landscape-level ecological restoration, the garden partners with the mining industry, a sector that has been thriving in Western Australia thanks to the rapid growth of the Chinese economy. Mining has a major impact on rare and threatened species in this global biodiversity hotspot and conservation has to involve working with the industry.

Initially, the garden developed a partnership with a mining company that was attempting to restore the hidden beard heath (*Leucopogon obtectus*), which was threatened by the mining operation. From this recovery work for a single species the programme grew into a larger research effort looking at the use of the plant family Ericaceae at mining restoration sites. Kings Park is now working with the sand quarrying industry to help with restoration of banksia woodlands. With mining industry support Kings Park scientists have been able to establish conservation techniques and management for rare and threatened species in non-mining areas.

## CLIMATE CHANGE

Responding to climate change is going to be a future challenge for the scientists at Kings Park and Botanic Garden. The garden itself provides an opportunity to monitor how plants respond to climatic conditions. In one particularly bad drought year around 10 per cent of all *Banksia, Eucalyptus* and *Allocasuarina* species succumbed to physiological stress thought to be caused by drought. The impacts of climate change on the flora of south-western Australia are expected to be significant given the concentration of naturally rare species in this cool, wet corner of the country. The already daunting plant conservation challenges will be raised to a new level. Research suggests, for example, that most of the endemic banksias will be unable to migrate to suitable habitat and may face extinction as a result of climate change. A similar scenario may be expected for many other endemic groups of plants. The Director of Kings Park and Botanic Garden, Mark Webb, takes both the threat and the challenge very seriously. 'Our work at Kings Park is not just about saving species for scientific interest, we have to be ready to use our botanical and horticultural skills to help maintain a vegetation cover in south-west Australia. We need to retain native vegetation as carbon sinks, restore vegetation working with the extractive industries and be ready to manage vegetation with different species compositions as a result of our changing climate.'

Webb sits on the Council of the Heads of Australian Botanic Gardens. Recently this body has produced a national strategy 'to promote and resource Australia's botanic gardens in their vital role as facilitators of climate change preparedness and adaptation'. As the report points out, not only are Australia's botanic gardens providing a safety net for wild plants through *ex situ* collections, they have unparalleled and unique knowledge about where plants grow naturally, how they can be propagated and grown outside their natural ranges and how this knowledge can be interpreted for different sectors of society. Botanic gardens are popular attractions in Australia, between them attracting a total of over 5 million visitors a year, most of whom vist more than once. They are therefore ideally positioned to explain and demonstrate the relationships between biodiversity conservation, the impacts of climate-change and sustainable living practices.

Carnaby's black cockatoo is endemic to south-west Australia. Nesting in eucalyptus woodlands, and feeding on the seeds of native and introduced plants and insect larvae, this species is threatened by the clearance of natural vegetation.

Measuring the photosynthetic capacity of rare plant species restricted to banded ironstones in Western Australia. Scientists at Kings Park and Botanic Garden carry out innovative research in native plant biology, the results of which underpin the conservation and ecological restoration of biodiversity in Western Australia and beyond.

Botanic gardens have been at the centre of plant exploration and cultivation for the past 500 years. It is only in the last 30 years or so that plant conservation has become a major role. This role must become increasingly important as the world's flora continues to be impoverished. At a time of rapid global change and ecological uncertainty botanical arks will need to scale up their storage capacity and prepare for ecological restoration on an unprecedented scale. Thirty years ago the main task was to document threatened plants. Armed with information, botanic gardens have since made considerable progress in conserving threatened species. But global climate change is changing the parameters completely and the world's plants continue to face countless other threats. Many people are calling for the true conservation potential of botanic gardens to be realized.

**Previous page:** Meadow flora in Mount Rainier National Park, Washington, USA.

Around the world deforestation and agricultural intensification continue to be major threats to biodiversity. On top of these threats is the all pervasive impact of climate change that will place pressure on the natural range and survival of populations of all wild plants. Species that are already rare will become rarer still. The existence of many species in the wild will be threatened because many are restricted in range, and because environments will change faster than most plant species can move to different areas through seed dipersal or adapt. Levels of species loss will be directly related to the extent to which we can limit global warming. Recent models based on a temperature rise of 2–3°C over the next 100 years suggest that up to 50 per cent of the 300,000 or so higher plant species will be threatened with extinction. It is hard to imagine the impact of such a loss of plant diversity.

The structure and function of ecosystems are determined by their component species, primarily plants. There is a real fear that climate change will cause the extinction of so-called keystone species, leading to ecological collapse and the death of many additional species. Healthy ecosystems provide fundamental life-support services. As well as providing tangible products, such

as timber, food and medicine, ecosystems stabilize landscapes, help to purify air and water, remove toxins from the environment, and mitigate floods. These essential indirect services are of immense value but have scarcely been quantified in economic terms.

The role of individual plant species within an ecosystem is usually little understood. With such abundance of plant diversity, some ecologists have raised the issue of species redundancy, suggesting that possibly ecosystems contain far more species than are necessary to carry out particular ecological processes. But the capacity of a functional, intact ecosystem to respond to changes in the environment may depend on these 'redundant' species if it turns out that they are better adapted to new environmental conditions.

As well as providing essential ecosystem services, plants also directly or indirectly support people, by providing food, medicine, timber, fibres and many other materials essential for our daily lives. In developing countries, 2.5 billion people depend on agriculture for their livelihoods and activities directly based on plants account for approximately 75 per cent to the national economies of many of these countries. Furthermore, the livelihoods of over a billion people

Seeds of the jojoba plant, *Simmondsia chinensis*, native to the deserts of Arizona, California and Mexico. These seeds were collected in Arizona by the Seeds of Success programme, which has been collecting and banking seed of wild populations of native plants since 2001.

depend on forest resources, while 80 per cent of all people rely on traditional medicine – largely based on plants – for their primary healthcare. Human welfare is closely linked with the diversity of plants. As we lose these resources the future for the rural poor is likely to become increasingly difficult in many parts of the world.

## ECOSYSTEMS OR EX-SITU COLLECTIONS?

The ecosystem approach adopted by the Convention on Biological Diversity promotes the conservation of sustainable ecosystems with all species and people as an integral part. Protecting forests, grasslands and wetlands will help to slow down the rate of climate change because these ecosystems act as vital carbon sinks absorbing carbon dioxide produced by transport systems, industry, agriculture and homes around the world. To support ecosystem conservation, species under threat need special attention, whether in their natural habitats, within cultivated landscapes or within botanic gardens. Plant species in ecosystems of low-lying islands, mountain areas, arctic areas and at the edges of continents are likely to be particularly badly hit by climate change because they will have nowhere to go to as temperatures increase and sea levels rise. Assisted migration is being proposed by some ecologists but moving individual species or groups of species is a complex undertaking. It makes sense to keep all options open by storing plants – particularly those most vulnerable to climate change, and those important to people – in *ex situ* collections.

Scientists accept that it may not be possible to save all plants in their natural habitats. Documented living collections, seed banks and DNA banks provide an alternative and supplementary approach to *in situ* efforts. While prioritizing conservation of rare and threatened plant species in their natural habitats continues to be important, scaling up seed banking efforts for all wild plant species is a logical first step in a changing climate where virtually all species are at risk.

*Ex situ* collections play a vital role in conserving plant diversity, not only as an insurance policy for the future, but also as a basis for restoration and reintroduction programmes. Around the world, botanic gardens have been working towards a target of 60 per cent of threatened plant species in accessible *ex situ* collections, preferably in the country of origin, with 10 per cent of these species included in recovery and restoration programmes.

protecting forests, grasslands and wetlands will help to slow down the rate of climate change

# XISHUANGBANNA TROPICAL BOTANICAL GARDEN

One botanic garden that is preparing to expand its role to tackle broad environmental problems is the Xishuangbanna Tropical Botanical Garden in Yunnan, China. Renowned for its exceptional beauty, this garden was established 50 years ago as a research centre to develop the production of rubber in China. Now rubber is grown extensively in the area around Xishuangbanna, replacing the biodiverse rainforests. With the demand for cars booming in China rubber production is increasing and rubber plantations are directly competing with rainforest conservation. Scientists at Xishuangbanna are also concerned that forest clearance is affecting the regional climate and that the plantations are affecting water supplies to local villages. New measures are being developed by the local government to strengthen the protection of natural forests and restore the buffer zones of national nature reserves. In a broader conservation effort, the Asian Development Bank is supporting a Biodiversity Corridors Initiative involving Xishuangbanna and China's neighbouring countries of Laos, Myanmar and Cambodia. Financial support is provided through microcredit to villages which establish new protection zones connecting the nature reserves. The potential of carbon trading and biodiversity offsets are being explored to help conserve the forests and reduce poverty in the region. Xishuangbanna Tropical Botanical Garden is stepping up its research role with a major new research centre and is expanding its role in influencing biodiversity policy. As Chen Jin, the director of the garden has said, 'The future of the botanical garden is directly linked to the future of the ecological health of Xishuangbanna. The day when the natural forests have disappeared and the rivers stop running will be the end of our tropical garden haven. We cannot afford to lose the forests that support the traditional culture and livelihoods of this region.'

Many botanic gardens around the world are using laboratory techniques and research to complement field work. Seed banking, germination trials and genetic research are all part of the move towards prioritizing plant conservation.

PLANNING FOR THE FUTURE

In recent years major progress has been made towards this target, and 30–40 per cent of globally threatened species are known to be included in *ex situ* conservation programmes. BGCI's PlantSearch database records over 12,000 globally threatened plant species among the over 80,000 species held in botanic garden collections. Many other locally threatened species are also grown by botanic gardens. The database records medicinal plants and wild crop relatives, as well as propagation techniques and species used in restoration programmes. Institutions holding living collections are beginning to assess their conservation value, and they are taking steps to ensure that they can act as true safety nets for rare species. Globally, BGCI is able to identify the gaps in *ex situ* conservation for selected groups of plants. A survey carried out in 2009 suggested that three quarters of all European threatened plant species are not yet secure in botanic garden collections or seed banks.

The situation is more precarious for tropical plant species. Whereas 90 per cent of the world's temperate trees are in cultivation only about 10 per cent of tropical trees are cultivated in any form of plantation or garden. That is why the work of botanic gardens such as the well-established Rio de Janeiro Botanic Garden or the newly established Tooro Botanical Garden in Uganda are so important.

Even where plants are grown by botanic gardens there is no room for complacency. There are questions about the genetic 'representativeness' of many *ex situ* collections, and thus their suitability for use in restoration and reintroduction programmes. These questions must be addressed and steps taken to ensure that garden collections are increasingly representative of the genetic diversity of wild populations. In the past the desire to have impressive collections of cacti, orchids or palms led to the Victorian 'stamp collecting' attitude of botanic gardens. This has now completely changed as curators recognize that quality and relevance rather than quantity is important. There is little point in the costly maintenance of greenhouses full of ancient, undocumented pot plants.

## SEED BANKS

Complementing *ex situ* living plant collections are a large number of seed banks, usually run by botanic gardens, operating at local, national and international levels. Within Spain, for example, the Andalusian seed bank at

# WORKING WITH LOCAL COMMUNITIES: VIETNAM'S PHARMACOPOEIA

Nearly a third of Vietnam's 1,000 plant species have a recorded medicinal use, yet as forests are being degraded or destroyed, many medicinal plants – immensely valuable to local communities – are becoming increasingly vulnerable. The Botanic Garden of the Hanoi University of Pharmacy, a very small garden working with extremely limited resources, has been documenting the ethnobotany of hill tribes over many years and working on sustainable livelihoods for these communities.

Ba Vi National Park is situated in a mountainous area about 50 kilometres north of Hanoi. It covers an area of around 7,400 hectares with a strictly protected core area of relatively intact tropical forest. This is surrounded by a larger buffer zone where crops such as cassava are grown and cattle are allowed to graze. About 45,000 people live in the buffer zone area, including people of the Kinh, Muong and Dzao groups. Dzao people migrated to the area in the 1920s bringing a tradition of shifting cultivation. For the past 40 years they have lived in settled communities but most Dzao families continue to rely somewhat on harvesting natural resources. They collect medicinal plants from the park and surrounding areas both for their own use and as a source of income.

The Botanic Garden of the Hanoi University of Pharmacy is working with Dzao traditional herbalists to conserve the threatened *Ardisia gigantifolia* and *Stephania dielsiana*. Both species offer an important source of income to sustain livelihoods of local communities. A permanent propagation facility has been established in a home garden in Tan Linh Commune, adjacent to the National Park. Plants grown here have been supplied to local herbalists for their own gardens and planted back in the forest. This type of approach, sharing responsibility for conservation between botanic gardens and local communities, is increasingly important across the world.

the Cordoba Botanic Garden stores seed for this particularly plant-rich part of the country. The Viera y Clavijo Botanic Garden on Gran Canaria has a major seed bank for the Canary Islands, Azores and Maderia. In China a major new seed bank was established in the Kunming Institute of Botany, Yunnan, in 2008 with the aim of storing seed from 4,000 threatened and endemic plant species by 2010. The Millennium Seed Bank project, created by Kew and with over 50 partners worldwide now stores seed from 20,000 plant species, including both common and threatened species mainly from the world's drylands. Looking ahead, project aims to collect and conserve 25 per cent of the entire world flora by 2020. Emphasis in collection programmes will shift to species from montane, coastal and island ecosystems – the most vulnerable to climate change. Seeds of species that are of greatest utility to people will also be collected, including: drought-resistant crop and forage species that sustain some of the world's poorest rural communities; salt tolerant and desert pioneer species that combat desertification; plants with potential as biofuels; and medicinal plants.

Long term storage of some seeds is still a challenge. Although seed from approximately 90 per cent of flowering plants can be dried and stored at temperatures below freezing, problems remain for plants with so-called recalcitrant seeds. Unfortunately many such species are those of the rainforests and other tropical ecosystems. The length of time that seed can be stored and retain its viability is also very variable. Short-lived seeds need to be germinated at appropriate intervals so that new seed can be produced and stored. Every collection that comes into the Millennium Seed Bank is tested, with around 10,000 germination tests every year. The germination protocols that result from these studies are very valuable for establishing effective conservation and reintroduction strategies.

## RESTORING WILD POPULATIONS

Restoration of threatened plant species in the wild has so far been relatively restricted. However, recovery of wild populations of depleted species will be increasingly important in repairing damaged ecosystems and restoring habitats at a landscape scale. Successful habitat restoration will require genetically diverse populations of species to be maintained in *ex situ* collections, and gardens will need more resources to propagate and cultivate endangered species. A restored ecosystem should have a full complement of species or the potential to attract those that are missing. Scientists believe that the best way to attract a wide array of animals back to restored habitats is to establish a full complement of plants at the site. Ecological restoration is likely to be a major challenge in the 21st century, with botanic gardens helping to restore threatened species in their natural habitats and rebuild plant communities. Fundamental to this is an understanding of the identification and distribution of plant species so that the ecological jigsaw pieces can be reassembled.

Good examples of ecological restoration can already be seen around the world. In the US, the Seeds of Success (SOS) is an ambitious, forward-looking programme that is working with the Millennium Seed Bank and

**good examples of ecological restoration can already be seen around the world**

Collecting seed from the saguaro cactus, an iconic species of south-western USA and northern Mexico. Individuals take about 150 years to reach their full size. Mature plants produce millions of seed each year but few seedlings survive in the wild. The seeds germinate easily in cultivation.

Seeds of Success volunteers taking material for the herbarium. When collecting seed, herbarium specimens are important references for accurate identification.

many other partners. The programme has been collecting and banking native seed from wild populations since 2001 to restore the American landscape. Almost one-third of land in the US is managed by the Federal Government and restoring land that has been damaged by fire or over-grazing is a very important part of that management. Peggy Olwell, the founder of SOS, explains that, 'We need to think on a big scale if we are to conserve and restore our threatened

PLANNING FOR THE FUTURE

native flora in the US. The commercial seed suppliers cannot provide the quantity and range of plants that we need for restoration. Botanic gardens have the skills to identify, collect, store and germinate seed and are essential partners in the programme.' Changing climatic conditions make the need for seed banking increasingly important. In 2007, for example, southern states experienced serious droughts, tornadoes hit the Midwest and wildfires raged in California. Peggy considers that 'native plant communities are essential for sustainable economic prosperity and the quality of life that comes from public lands. We cannot afford to lose our wild plant species and the potential they offer for adaptation to climate change.'

There remains a need for a coordinated global approach to seed banking, with the current seed bank partnerships developed into a worldwide network. The seed of agricultural food crop varieties are being stored underground in the Svalbard Global Seed Vault – a similar bold vision and approach may now be needed for the seed of wild plants.

## THE IMPORTANCE OF WORKING TOGETHER

Working together botanic gardens increase effectiveness as they exchange skills and expertise and learn from successful conservation activities. New gardens, such as the Nezahat Gökyiğit Botanic Garden in Istanbul, learn from the well-established gardens such as Kew and Edinburgh. Gardens with limited resources benefit from international collaboration, as seen in Madagascar. Partnerships and networks help to develop conservation strategies and practical programmes in most countries of the world. But there are gaps in the safety net. A few countries in the Middle East, Asia and Africa do not have a botanic garden and there are relatively few botanic gardens in some of the countries richest in plant diversity, such as Indonesia and Peru.

The problems that botanic gardens are able to tackle are global. But the key to success in saving the world's plants will be input from us all, as Professor Stephen Blackmore points out: 'The pressures humanity imposes on the earth are the cumulative effect of 6.6 billion people going about their daily lives and, while political initiatives and leadership play a vital part, only the decisions and changes of ordinary people and their families can lighten the burden we place on our planet. It is imperative that people understand the fundamental role of plants in maintaining the biosphere and nowhere is better than a botanic garden to gain this understanding.'

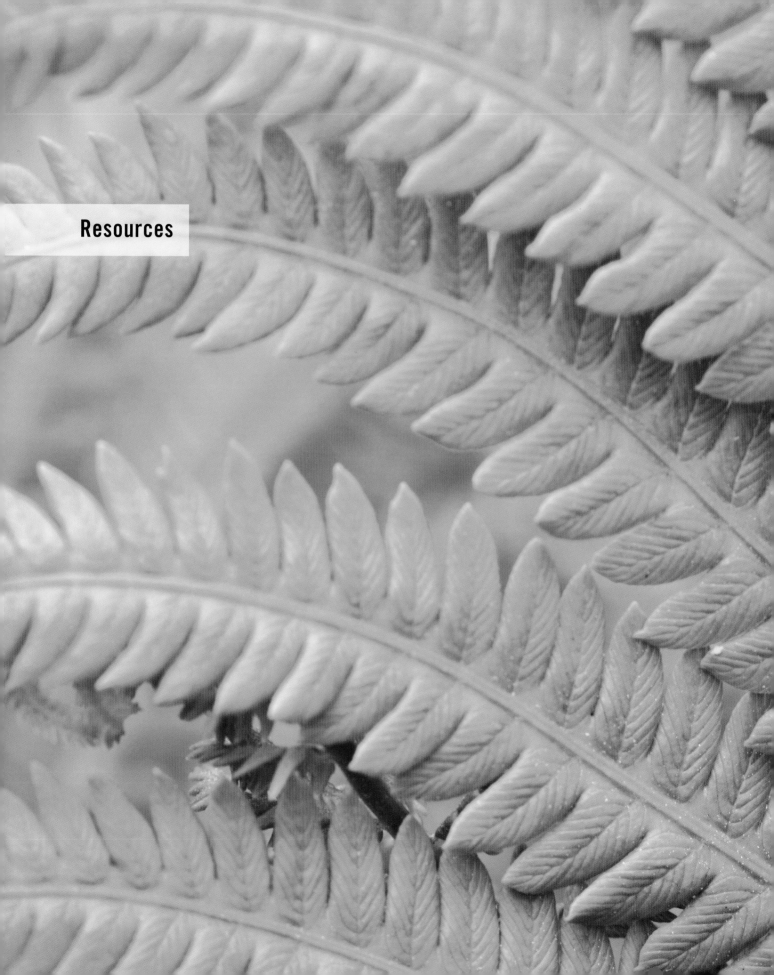

Resources

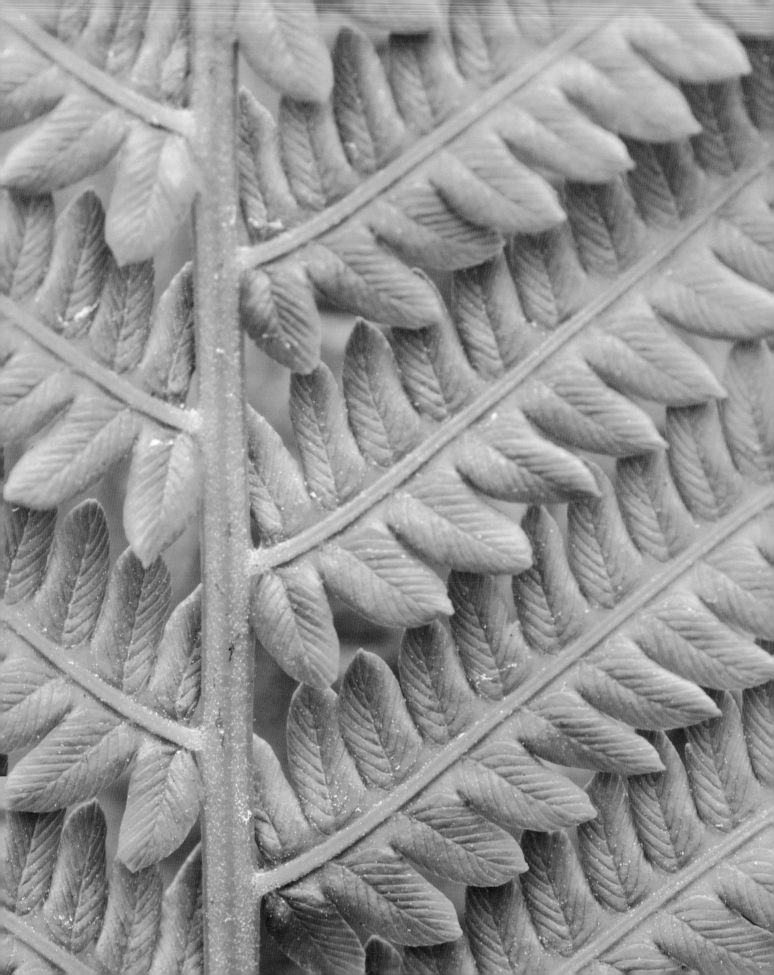

# Further reading

This short selection of books will introduce you to some of the key issues and ideas in plant conservation. Some titles are quite technical in their coverage, others are intended for a broader audience.

Blackmore, S. (2009). *Gardening the Earth. Gateways to a sustainable future.* Royal Botanic Garden, Edinburgh

Falk, D.A. Millar, C.I. and Olwell, M. (Ed.) (1996). *Restoring diversity. Strategies for reintroduction of endangered plants.* Island Press.

Guerrant, E.O., Havens, K. and Maunder, M. (Ed.) (2004). *Ex situ plant conservation. Supporting plant conservation in the wild.* Island Press, Washington.

Hamilton, A. and Hamilton, P. (2006). *Plant conservation. An ecosystem approach.* Earthscan, London, UK.

Hawkins, B., Sharrock, S. and Havens, K. (2008). *Plants and climate change:which future?* Botanic Gardens Conservation International, Richmond, UK.

Lynas, M. (2007). *Six Degrees. Our future on a hotter planet.* Fourth Estate.

Marinelli, J. (1994). *Plant.* Dorling Kindersley Ltd., London, UK.

Sharrock, S. and Jones, M. (2009). *Conserving Europe's threatened plants. Progress towards Target 8 of the Global Strategy for Plant Conservation.* Botanic Gardens Conservation International, Richmond, UK.

# Taking action!

Governments around the world have signed up to agreements on our behalf, but government alone will not solve the biodiversity crisis. Visit the websites given below to find out more about global conventions and voluntary organizations that are supporting plant conservation around the world, and find out how you can get involved.

### CONVENTION ON BIOLOGICAL DIVERSITY

The Convention on Biological Diversity (CBD) aims to conserve biodiversity, ensure the sustainable use of biodiversity and ensure the fair and equitable sharing of benefits arising from the use of genetic resources. The CBD Global Strategy for Plant Conservation (GSPC), agreed in April 2002, sets out specific targets for the conservation and sustainable use of plant biodiversity.
www.cbd.int

### CITES

The Convention on International Trade in Endangered Species of Wild Fauna & Flora (CITES) aims to ensure that international trade in wild animals and plants does not threaten their survival. Governments that comply with the Convention have their own national legislation to implement the Convention sometimes with stricter requirements on trade in wild plants and animals.
www.cites.org

### UN FRAMEWORK CONVENTION ON CLIMATE CHANGE

This Convention was developed for governments to collectively consider what can be done to reduce global

warming and to cope with temperature increases. The Kyoto Protocol is an addition to the Convention approved by a number of nations which has more powerful (and legally binding) measures.
www.unfccc.int

## UN CONVENTION TO COMBAT DESERTIFICATION
This Convention aims to tackle desertification through integrated approaches, emphasizing action to promote sustainable development at the community level.
www.unccd.int

## BGCI
Botanic Gardens Conservation International (BGCI) links botanic gardens in over 100 countries in a shared commitment to biodiversity conservation, sustainable use and environmental education. BGCI aims to mobilize botanic gardens and work with partners to secure plant diversity for the well-being of people and the planet.
www.bgci.org

## FAUNA AND FLORA INTERNATIONAL
Founded in 1903, Fauna and Flora International (FFI) is the world's oldest conservation organisation. FFI acts to conserve threatened species and ecosystems worldwide, choosing solutions that are sustainable, are based on sound science and take account of human needs.
www.fauna-flora.org

## GLOBAL TREES CAMPAIGN
The Global Trees Campaign is a joint initiative run by BGCI and FFI to save the world's most threatened trees and their habitats, through the provision of information, delivery of conservation action and support for sustainable use.
www.globaltrees.org

## PLANTLIFE INTERNATIONAL
Plantlife is a charity working to protect Britain's wild flowers and plants, fungi and lichens in the habitats in which they are found. Plantlife also works internationally on identification of Important Plant Areas, medicinal plant conservation, European plant conservation, and support for the GSPC.
www.plantlife.org

## THE INTERNATIONAL UNION FOR THE CONSERVATION OF NATURE (IUCN)
IUCN helps the world find pragmatic solutions to the most pressing environment and development challenges. It supports scientific research, manages field projects all over the world and brings people together to develop and implement policy, laws and best practice. The IUCN Red List provides the most authoritative information on the conservation status of wild species worldwide. The general aim of the system is to provide an explicit, objective framework for the classification of the broadest range of species according to their extinction risk
www.iucn.org

# Action at home

All life depends on plants and plant conservation needs sustained support. Some of the conservation actions described in this book are highly technical and need specialist equipment and skills. But we can all take action to help prevent plant extinctions and keep the planet green. As well as supporting the organizations listed in the previous pages, you can:

- Check which plant species in your garden are threatened in the wild using the online PlantSearch database at www.bgci.org.

- Enjoy wildflowers in their natural habitats and consider growing local species (grown from seed or cuttings, *not* dug from the wild) in your garden to encourage local wildlife.

- Visit your local botanic garden, become a 'friend' or volunteer. Visit botanic gardens on your travels to find out about the floras of other countries. To find details of botanic gardens and their locations visit the GardenSearch database at www.bgci.org.

- Be a green gardener: avoid using peat; use chemicals sparingly, if at all; be water-conscious – collect rainwater and use 'grey water' from household washing; make compost; recycle.

- Avoid planting invasive species in your garden. Visit www.issg.org/database for more information. In the UK, visit www.plantlife.org.uk.

- Do not buy cacti, orchids or other ornamental plants that have been harvested from the wild.

- Buy sustainably produced timber to reduce logging damage to forest ecosystems that are rich in plant diversity.

- Take steps to reduce your carbon footprint! Climate change is a major threat to plant diversity worldwide.

# Acknowledgements

Many botanic garden friends and colleagues have contributed ideas, suggestions and information for this book and I am grateful to them all. Particular thanks are due to the following for their help, for dealing with my persistent requests for information, and for assistance with sourcing photographs:

Hamish Adamson, Peter Baxter, Chris Birkinshaw, Professor Stephen Blackmore, Fiona Bradley, Dr David Bramwell, Dr Stéphane Buord, Alec Burney, Dr David Burney, Dr Thomas Busch, Andy Byfield, Dr Javier Caballero, Carly Cowell, Dr Chen Jin, Maite Delmas, Dr Kingsley Dixon, Dr Mike Fay, Gina Fullerlove, Martin Gardner, Dr Kay Havens, Dr Héctor M. Hernández, Hendrian Hendrian, Anthony Hitchcock, Professor Stephen Hopper, Professor Huang Hongwen, Hu Huabin, Margaret Johnson, Dr Adil Güner, Dr Andrea Tietmeyer Kramer, Olivia Kwong, Leeann Lavin, Chris Leon, Dr Carl Lewis, Dr Cornelia Loehne, Professor David Mabberley, Dr Joyce Maschinski, Dr Mike Maunder, Scot Medbury, Quentin Meunier, John Wasswa Mulumba, Alice Notten, David Nkwanga, Peggy Olwell, Dr Gerald Parolly, Tania Sampaio Pereira, Jennifer Possley, Fano Rajaonary, Professor Peter Raven, Godfrey Ruyonga, Dr George Schatz, Mustaid Siregar, Albert Dieter Stevens, Dr Nigel Taylor, Mark Webb, Xiangying Wen, Didik Widyatmoko, Christopher Willis, Dr Peter Wyse Jackson and Dr Daniela Zappi. Thanks are also due to Barbara Bridge, George Hackford and Fiona Wild who helped compile draft material for the book and to Joachim Gratzfeld, Bian Tan and Suzanne Sharrock who advised on sections of it. I am also very grateful to Beth Rothschild for taking the time to read through the manuscript and prepare such a thoughtful foreword.

The generosity of botanic gardens that have made freely available photographs for use by BGCI in preparation of this book is gratefully acknowledged. The individuals credited are all thanked sincerely for providing their images of plants and gardens. BGCI staff members have also made available photographs they have taken as a contribution to BGCI's picture library. Liz Smith assisted with picture research and efficiently organized and collated the pictures used throughout the book.

Thank you again to all who have helped.

## Picture credits

The author and publisher have made every effort to trace and credit the copyright holders of the images which appear in this book and will be pleased to rectify any omissions or errors at the soonest opportunity.

Front cover: *Trillium*, Chicago Botanic Garden. © Robin Carlson/CBG; Back cover: *Aloe dichotoma*, Karoo Desert National Botanic Garden. © Lize Wolfaardt Brooklyn Botanic Garden, 19; Meneerke Bloem/Wikimedia Commons, 67; BGCI; 68 Alec Burney, 117, 120, 122; Robin Carlson/CBG, 4, 6; Peter Clarke/RBGE, 41, 43(m), 45(l); Robin Carlson, 98, 101, 102, 104, 106, 109; Liu Chunfang, 144; Han Chunyan, 150(r); Christian Claude, 184(r); Todd Erickson, 215; Martin Gardner/RBGE, 43(b), 45(r); Adil Güner 72 75, 76(l), 76(r), 77, 78, 82; Joie Goodman, 92(l), 92(r), 93; P. Goltra/ NTBG, 118, 121; Peter Hollingsworth/RBGE, 45(t); Ingo Hass/BGBM, 62; Jacaranda Photography, 21, 206, 217, 224; Margaret Johnson, 79(b), 80, 81; Yang Kerning, 150, 151; Barbara Knott, 210, 216; Paul Little/RBGKew, 30; P. La Croix/CBN Brest, 53; Zhang Lehua, 156(t and b); Mickael Mady/CBN Brest, 11; A. McRobb/RBGKew, 25, 26, 28, 32; Chris Migliaccio, 90 (l), 90(r); Quentin Meunier, 172, 173(l), 173(r), 175, 176, 177(t), 177(b); Naturepl.com: Jack Dykinga 124, David Fleetham 110, Nick Garbutt 185, Konstantin Mikhailov 70, Pete Oxford 178, Robert Valentic 209, Dave Watts 141; NBBG, 79(t); Sara Oldfield, 17; photolibrary.com: Nigel Dennis 204, GTW 134, GTW 134, Mark Hamblin 46, 218, David Kirkland, 163, Peter Lilja 22, Roel Loopus 211, Mauritius 65, OSF 212, Richard Parkwood/OSF 8, Nigel Pavitt 168; Jennifer Possley, 84–85, 87, 88(l); Zey Qingwen, 152; Tania Sampaio Pereira, 137, 139; Ouyang Pei, 147, 148; Zhang Qianmei, 154(r) RBGKew, 35; L. Ruellan/CBN Brest, 50, 55, 56, 57(l), 57(r), 58, 59(l), 59(r), 60; Julissa Roncal, 91, 95, 97; Singapore Botanic Garden, 18; Zeng Songjun, 154(l); SANBI: 197, 198, Carly Cowell 201, 203, Anthony Hitchcock 193, 204; Alice Notten 195, 196; Lize Wolfaardt 199, Phakamari Xaba 202; Kjell Sandred, 190; SOS: Desert Botanical Garden 221, 228(l), Student Conservation Association, 228(r); Bian Tan/BGCI, 13; Patti Vitt, 105(l), 105(r); Lynsey Wilson/RBGE, 2, 14, 36, 39, 43(t), 49; Debbie White/RBGE, 40, 42; Kristie Wendelberger, 88(r); Sam Wright, 89(t and b), 94, 96

First published in North America by The MIT Press.
Published in the United Kingdom by New Holland Publishers (UK) Ltd

Text copyright © Sara Oldfield/BGCI
Copyright © 2010 in photographs: see photo credits
Copyright © 2010 New Holland Publishers (UK) Ltd

Sara Oldfield has asserted her moral right to be identified as the author of this work.

MIT Press books may be purchased at special quantity discounts for business or sales promotional use. For information, please e-mail special_sales@mitpress.mit.edu or write to Special Sales Department, The MIT Press, 55 Hayward Street, Cambridge, MA 02142.

Library of Congress Cataloging-in-Publication Data
Oldfield, Sara.
Botanic gardens : modern-day arks / Sara Oldfield.
  p.   cm.
Includes index.
ISBN 978-0-262-01516-5 (hardcover : alk. paper)
1. Botanical gardens. 2. Plant conservation. I. Title.
QK71.O43   2010
580.7'3—dc22
                                    2010017196

Printed and bound by TWP in Singapore